ENGINEERING FOR ARCHITECTURE

Architectural Record Books

Affordable Houses
Apartments, Townhouses and Condominiums, 2/e
The Architectural Record Book of Vacation Houses, 2/e
Buildings for Commerce and Industry
Buildings for the Arts
Engineering for Architecture
Great Houses for View Sites, Beach Sites, Sites in the Woods, Meadow Sites, Small Sites, Sloping Sites, Steep Sites and Flat Sites
Hospitals and Health Care Facilities, 2/e
Houses Architects Design for Themselves
Houses of the West
Institutional Buildings
Interior Spaces Designed by Architects
Office Building Design, 2/e
Places for People: Hotels, Motels, Restaurants, Bars, Clubs, Community Recreation Facilities, Camps, Parks, Plazas, Playgrounds
Public, Municipal and Community Buildings
Religious Buildings
Recycling Buildings: Renovations, Remodelings, Restorations and Reuses
Techniques of Successful Practice, 2/e
A Treasury of Contemporary Houses

Architectural Record Series Books

Ayers: Specifications for Architecture, Engineering and Construction
Feldman: Building Design for Maintainability
Heery: Time, Cost and Architecture
Heimsath: Behavioral Architecture
Hopf: Designer's Guide to OSHA
Portman and Barnett: The Architect As Developer

ENGINEERING FOR ARCHITECTURE

by Robert E. Fischer

An Architectural Record Book
McGraw-Hill Book Company

New York
St. Louis
San Francisco
Auckland
Bogotá
Hamburg
Johannesburg
London
Madrid
Mexico
Montreal
New Delhi
Panama
Paris
São Paulo
Singapore
Sydney
Tokyo
Toronto

The editors for this book were
Jeremy Robinson and Patricia Markert.

The designer was Irving Weksler.

The production supervisors were Elizabeth Dineen and Sara L. Fliess

The book was set in Palatino by Chester Graphics Service, Inc.
and in Optima by Jemet, Inc.

Printed and bound by Halliday Lithograph Corporation.

Library of Congress Cataloging in Publication Data
Fischer, Robert E.
    Engineering for architecture.

    "An Architectural record book."
    Includes index.
    1. Building.   2. Structural engineering.
I. Architectural record.   II. Title.
TH845.F57   690   79-19891
ISBN 0-07-002353-0
234567890 HD HD 89876543210

CONTENTS

(continued)

CONTENTS
(continued)

Introduction

The main purpose of this book is to portray the inventiveness and resourcefulness of engineers working in the support of architecture. The book comprises a collection of innovative concepts in structures; in prefabrication of building shells and factory packaging of mechanical subsystems; in energy-conserving building designs and hvac systems and equipment; in architecturally significant solar heating/cooling systems; and in energy-conserving, flexible, architecturally appropriate lighting systems that provide good task illumination while also enhancing the visual environment.

The case examples cover a spectrum of building types and sizes, providing thought-provoking ideas for architects in all sizes of offices and areas of endeavor. Some of the more spectacular examples are suited only to large-scale structures—such as the "bundled-tube" concept used in the 110-story Sears Tower, or the long-span truss and air-supported roofs for huge sports stadia. Even so, the thought processes involved in all of the examples are instructive in one way or another, serving to outline important basic principles and to explain reproducible technique.

Though engineering can be form-giving in architecture, we see in the last decade a more natural incorporation of engineering as form, rather than the forced "showing off" of technology just because it was new and available and—as many of us thought—connoted progress. All of us are guilty of a little muscle-flexing now and then, and some occasions may call for it, but it is safe to say that engineers, just as much as architects, are more comfortable with approaches that mesh with over-all program requirements and economics. In this context it is rewarding to see high-rise structures being done with finesse and cost savings while solving architectural design problems—and offering form as an additional benefit (see Chapter 1). And long-span fabric structures saving money over more conventional techniques, letting daylight in, and providing mathematically and architecturally sound shapes (Chapter 2).

The construction industry is much maligned by outsiders for not moving faster on technological fronts. But this can be refuted on many counts, and one of these is the way this country has been utilizing prefabrication (Chapter 3). We discovered early on the problems in trying to apply European technology in systems building directly to the United States building industry. We learned how to make modifications, exploit domestic construction expertise and management, and make the approach viable here.

Several trends in the mechanical systems field have picked up momentum: 1) recapturing internal building energy that was once thrown away, 2) delivering heating and cooling effects with less fan energy, 3) combining mechanical components into larger and larger factory-built packages, and 4) exploring the practical side of solar energy. A slice of these technologies is presented in Chapter 5.

The engineering discipline that has changed most radically in the shortest time and is having the biggest impact on energy consumption in commercial and institutional buildings is lighting. The focus is on providing just the right amount of light in the right places to provide light to see and light to see by (Chapter 6). Because of the accent on non-uniformity and flexibility, ceilings, and hence rooms, are acquiring a different appearance. There is greater emphasis on the different lighting requirements of different tasks. And, one hopes, lighting will be designed with greater awareness of what it can do to mold space and create mood.

The attitude of progressive engineers tackling the gamut of problems presented to them by progressive architects is perhaps best summed up by structural engineer Michael Barrett: "Sure we like the hard jobs, but even on ordinary jobs we have fun coming up with imaginative ways to save money, or speed construction, or do better justice to the architect's design" (Chapter 4, "Structural engineering collaboration in architecture"). Furthermore, the best way for engineers and architects to get work done is to do a good job. Says Barrett: "When you get down to it, the only way to get business is to do a superlative job. We all seem to grow by taking on challenges, learning something new. We like to think we add to the profession by doing it—and we have our share of fun."
Throughout the book we see engineers excelling as problem solvers, and responsive architects collaborating in the process.

CHAPTER 1:

Structures–The Tall Ones

In designing the structures for very tall buildings, the engineer's problems of channeling gravity loads to the ground and providing resistance to wind loads are much intensified, and of these two, the wind problem is the most difficult.

Gravity loads are challenging to handle when there are only a few widely-spaced columns at the building's base, or when some columns in a building's exterior logically should be larger than others—as in some types of concrete construction. This can lead to an interesting textural expression in building facades.

Wind loads are challenging to handle when engineers cut down on the size and weight of structural members, and limit sway. This limited sway minimizes discomfort for occupants and reduces potential damage of cracked walls and partitions. Because the traditional method of providing resistance to wind—truss-braced cores with conventional beam-and-column framing for perimeter walls—requires too much material after about 20-30 stories, engineers drew on their ingenuity to invent "belt" trusses that engage core and exterior columns and throw more wind load into exterior columns; and the cantilevered tube, with closely spaced columns that take all the wind load.

Because the lightweight buildings of recent years don't have the heavy mass of stone masonry to give them the natural damping that diminished sway effects, engineers have had to discover new techniques to accomplish the same thing—and this has led to the development of the tuned mass damper (Citicorp Center) and the utilization of specially-developed plastics for dissipating wind energy as heat (World Trade Center).

New skyscraper forms from Chicago

The Chicago office of Skidmore, Owings & Merrill, located in the city where the sky-scraper originated, has taken this architec-tural form to new-found heights, and in the process has produced architectural forms that express the rationality of the structural systems, and that exploit their planning potentialities.

Early skeleton frames still carried heavy loads of masonry, though the frames were designed to carry gravity and wind loads to the foundation. Wind load was not much of a problem then, but it became one when buildings shed their heavy masonry skins (which added damping), and the structures had to do all the work.

When buildings are not very high, rigidly connected beams and columns can carry the wind. But the post-and-beam ap-proach becomes inefficient after about 20 stories. Other systems that supplant post-and-beam also reach limits in efficiency as they reach greater heights.

The result is that as structures have thrust higher—20, 40, 60, 100, 110 stories—new families of structural systems have evolved, each suitable for given ranges of heights in steel, concrete, or their combina-tion.

What these families of systems are can be seen most clearly in the work of the Chicago office of SOM over the past 15 years. Their achievements in the skyscraper genre stem from a unique combination of individuals, plus the emphasis put on very early collaboration between engineers and architects. And it can do so because of hav-ing both strong engineering and architecture inputs in-house. Discussions start when only the building program is more or less known—and nothing has even been sketched. Archi-tecture and engineering are then discussed together to try to synthesize them into a coherent building form.

The buildings and structures that then emerge from the SOM office are a result not only of this philosophy, but also of the types of people involved: the structural engineer has to be somewhat of an architect, and the architect somewhat of an engineer.

A very close interaction between their thoughts must occur.

A case in point is Sears Tower. The bundled tube structural concept Fazlur Khan developed meshed with design partner Bruce Graham's search for a shape that could gradually drop off floor areas as the building rose higher, to give the different sizes of floors the client wanted.

Khan feels that teaching is a very im-portant part of his professional life—the work with students helping to stimulate new ideas and concepts, as well as to think them through. He proudly points to the high competence-level of engineers in his de-partment—attributing a high efficiency of output, in conceptual and technical terms, to this fact. He believes the engineer's role, as the architect's, is to make solutions as simple and direct as possible. That out of simple logic and simple structural solutions, good, and great, architectural forms can develop.

The architect seeks a flexible,
uncluttered plan, and an economic height;
the engineer seeks the simplest way
to bring loads down to the ground

When the skyscrapers really began to go "up" in numbers and height in Chicago about 15 years ago, significant changes in structural design approaches began to emerge from the office of Skidmore, Owings & Merrill, there. Even before that in 1958, the firm produced a bold, husky expression for Inland Steel's 60-ft-clear-span rigid frame of 19 stories. Three years later saw the 20-story Hartford Building which gave a clear, strong expression of a concrete flat plate design in 22-ft-square bays. Then in 1964, SOM stretched the bay sizes to 36 ft in the 19-story BMA building in Kansas City. The rigid-frame steel structure is welded, and high-strength steel was used in the 36-ft-long girders. Projecting in front of the glass, the structure is one of the clearest expressions of a steel rigid frame.

In a frame structure, the total lateral drift caused by wind is due to primary factors: 1) bending moments in the girders (65 per cent of the total), and bending moments in the columns (15 per cent); and 2) axial stresses due to the overturning moment, resulting in column shortening and lengthening (20 per cent). Obviously drift has to be controlled to prevent undue wracking of partitions and windows, and to avoid building movement being unpleasantly perceptible to the occupants.

Fazlur Khan, partner and chief structural engineer of SOM, Chicago, has demonstrated in a number of technical papers that the structural performance of a rigid frame can be improved when a vertical shear truss or shear wall is combined with it. The drawings below show that the frame tends to pull back the shear truss or wall in the upper portion of the building, and push it forward in the lower portion. As a result, the frame is more effective in the upper portion where the wind shears are less (they go from zero at the top and build up to maximum at the base), and the shear wall or truss carries most of the shear in the lower portion of the building, where the frame cannot afford to carry high lateral load. This construction in which the shear truss interacts with the frame has been used in a number of buildings in the 40-story range.

For example in the Chicago Civic Center (C. F. Murphy and SOM, associated architects), the upper half of the building is a pure rigid frame construction, while the lower half is a shear truss-interaction structure. When a rigid frame is combined with a shear truss, the lateral sway is frequently reduced to 50 per cent of that if the truss had been used alone, and, further, the distortion of the floors is less.

This same approach works in concrete, too, with the "shear truss" being replaced by a "shear wall." SOM's example here is the 38-story Brunswick building in downtown Chicago. Finished in 1962, it was one of the first major-size buildings in Chicago

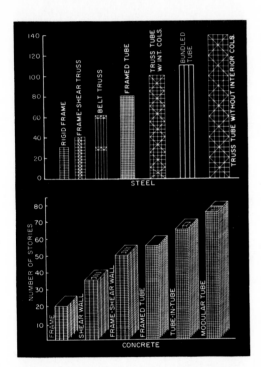

SHEAR WRACKING COMPONENT | CANTILEVER COMPONENT (COL. SHORTENING) | FREE FRAME | FREE TRUSS | COMBINATION FRAME & TRUSS | SHEAR TRUSS-FRAME INTERACTION

The taller buildings become, the stiffer they need to be to resist wind economically. The evolution of structures, including new concepts, to do this is shown, left. Low buildings up to 20 stories use rigid frames to limit sway, with wracking accounting for about 90 per cent of it. The 19-story BMA building (below) is a classic expression of a steel rigid frame. Because rigid frames are limber to some extent, they are inefficient for taller buildings. A first step to improve them is to add a shear truss (see above) which increases stiffness of the frame.

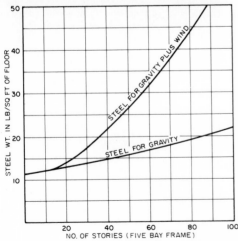

Up to 20 stories, the structures of steel frame buildings pay no penalty for wind resistance, but above that height the extra steel required for the frame to withstand wind increases radically, compared to that for gravity load.

to be built after the Prudential building. The program called for deeper space than usual—a 38-ft span from perimeter to core. In plan there is a 38-ft free span, a 38-ft corridor, and then another 38-ft free span.

At first SOM's engineers thought that the structure would be designed so that the core's shear walls would carry all the wind load, while the columns would carry only gravity load. But because of the long clear spans, columns had to be closer together than ordinarily—in this case 9 ft 4 in. apart, which was double the building module, and equal to the size of a "minimum" office. Obviously the columns of the exterior wall would not just "sit there." Because the frame was concrete, the columns and beams had a natural continuity. In essence, then, the building had shear wall-frame interaction. As a matter of fact, the engineers determined that with the building designed, the shear walls alone would allow the building to drift 13 inches with the strongest wind. But combining the shear walls with rigid frame action, the drift would be reduced to only 3 inches.

Concrete was chosen because at that time it was on the order of $1 per square foot cheaper than steel. Further, the closely spaced columns and the spandrel beams provided a natural frame for the windows.

In order to create adequate spaces for entry to the building, the individual loads of the closely-spaced columns had to be picked up by a huge transfer girder, 24-ft high and 8-ft deep, supported by 7- by 7-ft columns spaced 56 ft apart. Though the girder was huge, it served well the use of caisson-to-rock foundations, and the space behind it was used for location of the boiler and mechanical equipment.

A one-way joist type of slab was used between the exterior columns and the core, and this led naturally to a two-way waffle system at the corners. Because columns at the edge of the waffle are loaded more than the others, these columns were made deeper. Water riser details were manipu-lated at the other columns to match the two deeper ones near the corners. In later SOM buildings, the columns have been allowed to project on the outside, forming part of the visual expression.

For steel buildings in the 50-story range, the efficiency of the structure has been increased by tying the exterior columns to the core with belt trusses

It was pointed out earlier that the rigid frame structure, with bays of fair size, is inefficient because of the bending in the columns and beams. This can be improved upon, however, by connecting all exterior columns to the interior shear truss by means of belt trusses, which can increase the stiffness of the structure by about 30 per cent. When the core tries to bend under wind load, the belt truss, acting like a lever arm, throws direct axial stresses into the columns—compression on one side, and tension on the other. (An outrigger truss of this type was used in the U.S. Steel

The shear truss is more effective at lower floors, where the loading effect of the wind is largest, because effect of cantilever bending there is least (see diagrams across page). In Chicago's Civic Center, above, wind is resisted by a shear truss, rigid-frame combination in the lower floors, and by the rigid frame, alone, in upper floors. A similar kind of structural behavior is obtained in concrete by using an exterior rigid frame working together with concrete shear walls in the core. This approach was used in 1962 in the Brunswick building in Chicago by SOM (right). The floor framing is a one-way joist system, except for the corners which are two-way waffle slabs. Columns at the transition between the one-way system and the waffle slab are larger because of carrying more waffle weight.

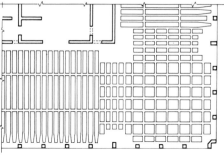

building and interior lateral trusses are being used in the I.D.S. building in Minneapolis—designed by other engineers).

Fazlur Khan first proposed belt trusses for the BHP Headquarters building in Melbourne, Australia, designing the structure for it. Comparative deflection curves for that building, with and without the belt truss system, are shown below. Obviously, the steel belt truss system at mid-height of the building contributes substantially to the stiffness of the building, as does the one at top.

A similar system has been employed in the 42-story First Wisconsin Center in Milwaukee by SOM. Here, not only are belt trusses used at mid-height and at the top, but a truss at the bottom is used as a transition member to collect column loads.

Shear wall design long has been a means for stiffening apartment buildings up to 30 stories and office buildings up to 20 stories or so. Studies for SOM projects have shown that over 30 stories, lateral sway as well as wind stresses begin to control the design, and structural elements designed only for gravity loads need to be made larger for stiffness and strength.

All approaches for optimizing tall skyscrapers have one thing in common: increasing the rigidity of the structure so it performs as a cantilevered tube

The floor plan of an apartment building wants to be more flexible than that of an office building; further the core is smaller, so it is better from these standpoints if the exterior walls alone could do the work in resisting wind, and that the shear walls be omitted. Maximum efficiency for lateral strength and stiffness, using the exterior wall alone as the wind-resisting element, can be achieved by making all column elements connected to each other in such a way that the entire building acts as a hollow tube cantilevering out of the ground.

Such a scheme was conceived in 1961 for the 43-story DeWitt Chestnut apartment building on Chicago's north side. The structure was thought of as a cantilevered tube with holes punched in it for windows, with smaller holes in the lower part and larger holes at the top because forces are less in the upper part. This tube was achieved in practice by having closely spaced columns (5 ft 6 in. centers) acting together with the spandrel beams, and this system is called the "framed tube."

The framed tube has limitations when used in buildings over 400 ft high because although the system looks like a tube, the two faces parallel to the wind act like a multi-bay rigid frame. As a result, the bending moments in the columns and edge beams become the controlling factor in unusually tall buildings. Further, of the total lateral sway, only about 25 per cent is due to column shortening caused by the cantilever action of the framed tube; 75 percent is caused by frame wracking. The phenomenon is known as shear lag, and is shown at the bottom of page 7. Ideally the shear transfers should be a linear rela-

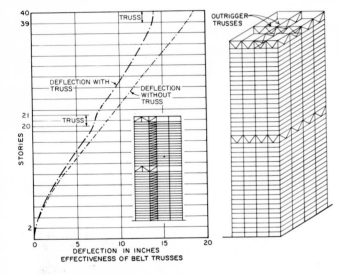

EFFECTIVENESS OF BELT TRUSSES

Above 40 stories the shear-truss, rigid-frame combination requires more and more steel for wind load. The effectiveness can be increased, however, by tying the shear truss to the exterior columns with belt trusses. The belt trusses, working as lever arms, throw direct axial stresses into the exterior columns. When the shear truss tries to bend, the exterior rows of columns act as struts to resist this movement. These belt trusses can be used not only at the top of the building, but midsection as well, increasing the stiffness of the building by 30 per cent. This approach has been used by SOM for the 42-story First Wisconsin Center in Milwaukee shown in the model photo below.

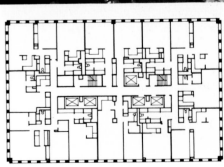

Rigid frames in concrete can be made more effective if the columns are spaced closely enough together so that the exterior structure works like a cantilevered tube when wind-loaded. The approach is especially favorable for apartment buildings, such as SOM's DeWitt Chestnut, in which core areas are small and planning flexibility is very desirable.

tionship; i.e., stresses in the building faces parallel to the wind should be direct tensions and compressions. But because of wracking of the frame, bending occurs, and columns at the corners of the building have to take more than their share of the load, while columns in between do less work than they ought to—so efficiency is reduced to the extent that beams and columns are limber, and consequently to the extent the frame wracks.

Framed tubes suffer from a problem called shear lag because the columns and beams bend when the wind blows. One remedy: stiffen the wall with diagonals

Exterior wall frames can be made stiffer and more rigid to mitigate wracking, however (and thus so-called shear lag). One method is to use diagonals in the wall, and, of course, the most striking example of this approach is the 100-story John Hancock building. The system used is the optimized column-diagonal truss tube. Ob-

viously the most effective tube action would be obtained by eliminating vertical columns and replacing them with closely spaced diagonals in both directions. But this not only presents problems in terms of window details and the large number of joints between diagonals, but the diagonals are less efficient than vertical columns in bringing gravity loads down to the ground. The column-diagonal tube, therefore, is an efficient compromise. The exterior columns have normal spacing, but are made to act together as a tube by the widely spaced diagonals. Except at levels where diagonals meet at corners of the building, the spandrels will resist the internal forces between columns and diagonals, but at these points it is necessary to provide a large tie spandrel to limit the horizontal stretching of the floors, and to make the diagonals function more efficiently as inclined columns, and as primary load-distribution members.

A similar approach can be worked out in concrete, as well. With the rigid tube

type of design it should be possible for concrete buildings to go 70, 80, even 100 stories. In contrast, with conventional beam and column framing, the practical height limit is on the order of 20 stories.

One way the rigidity can be achieved is with the column-diagonal approach. The diagonals can be created by filling in what normally would be windows in a diagonal pattern. With a rectangular building the diagonals will not cross on the wider faces, but they need to on the narrower faces for efficient transfer of wind load. Symmetry occurs about the corners, but not the faces of the building.

Still another approach in concrete that produces nearly 100 per cent rigidity is the interior bracing of the tube. A wall grid of closely-spaced columns is in effect "glued" to cross shear walls, so that the wall grid acts like the "flange" of a huge "beam," and shear walls act like "webs." Shear lag would be minimized, and stresses in the walls would be primarily axial.

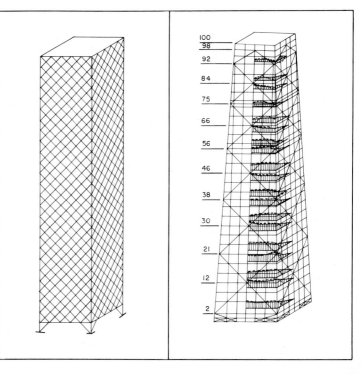

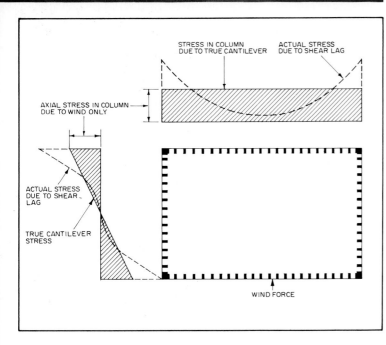

"Framed tube" is the designation given to structures that use closely-spaced columns in the exterior wall for wind load. But efficiency of framed tubes drops off in taller buildings (about 50 stories in concrete, 80 stories in steel). Ideally, columns and beams of a rigid frame would be infinitely stiff. But because these elements bend, a phenomenon occurs called, "shear lag," illustrated at left. Columns near corners do more work than they should; the others less. Shear lag can be greatly reduced by stiffening up the exterior; the stiffest means would be to replace vertical columns with diagonals. A more optimum approach from standpoints of overall efficiency and practicalness is to combine columns and diagonals as in the 100-story John Hancock building.

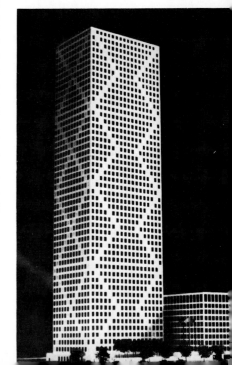

Efficiency of the framed tube can be improved if the interior core is also a tube, or if the exterior walls are braced by cross stiffeners

This scheme was used in a hypothetical 92-story apartment building by one of Fazlur Khan's students at Illinois Institute of Technology. For the system to work the shear walls have to be relatively continuous. With apartments having only an 8 ft 8 in. floor to floor height, openings in the shear wall for corridors could not be all in a vertical line because the shear wall "web" would be too weak. The problem is solved by using two different floor plans for alternate floors so that corridors, and thus openings, are staggered floor-to-floor.

A model was built in plastic, load tested, and found to be amazingly efficient. The system appears so simple and efficient that its actual application in an ultra-high rise building seems inevitable one of these days.

It has been shown that a concrete rigid frame and shear walls could interact to improve the performance of both, as in the Brunswick building. Going a step further, if the exterior wall is comprised of closely spaced columns so that it performs as a tube, and shear walls at the core also work as a perforated tube, then the structure becomes a "tube within a tube." The framed tube and shear wall-frame interaction concepts have been combined, and Fazlur Khan used this approach with the 52-story One Shell Plaza building in Houston. The building, at 715 ft, was, when built six years ago, the world's tallest reinforced concrete building, and the tube-in-tube concept made it possible at the unit price of a 35-story shear wall structure. The entire system is so efficient that all columns, shear walls and floors need be sized only for gravity loads. As with Brunswick, one-way joist system was used, in this case spanning 40 ft from exterior to core; columns were spaced 6 ft apart. The corners are a two-

way waffle slab, and again, as in Brunswick, exterior columns near the corners of the waffle are more heavily loaded by gravity than the other columns. But in contrast to Brunswick, these columns get gradually deeper, the additional depth is allowed to project out from the face in the building, and this gravity-load-carrying picture is expressed "plastically" in the building's exterior. In further contrast to Brunswick, the base of the building is pierced by much smaller openings.

Such a tall building would not have been possible in Houston—because of poor soil conditions—if the structural engineers had not searched out the possibilities of high-strength lightweight concrete in the range of 6,000psi for the entire structure. With conventional stone concrete, 35 stories would have been about the limit.

Further, the plan shape was changed from an original 120 by 240 ft (a tremendous "sail" area for Houston's 40 lb per sq ft wind load) to 192 by 132 ft—a ratio of

Ezra Stoller © ESTO photos

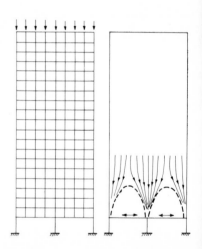

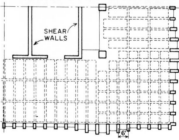

The concrete framed tube can be improved by making a structural tube out of the shear walls. The approach—called "tube-in-tube" was used for the 52-story One Shell Plaza. A picture of the increasing gravity loads in the columns next to the waffle slab can be seen in the undulated exterior. Increasing sophistication in collection of gravity loads of exterior columns is manifested in both One Shell Plaza, left and above, and in Rochester's Marine Midland bank, right. In the former, a massive base is pierced for access. In Marine Midland, the structure grows like a tree at the base.

1:1.45 rather than 1:2. The foundation consists of a concrete mat sitting 60 ft below ground; it is over 8-ft deep and projects out 20 ft from the perimeter of the superstructure.

Funneling the gravity loads of closely spaced columns into wider-spaced columns at the base in the structural design also makes possible new visual expressions
Collecting the columnar gravity loads by means of a deep transfer girder is rather a brute-force approach, inasmuch as the girder has to work in inefficient post-and-beam fashion. So, more recently, SOM's architects and engineers have taken a closer look at the load flow in a rigid wall of closely-spaced columns, supported by widely-spaced columns at the base. The natural load flow is for columns to gradually shed their load toward the base columns. The wall, in effect, actually works as an arch. Recognizing this, SOM has done several buildings in which columns and spandrel

beams grow larger as they approach the base columns. The most sophisticated of these buildings so far is the Marine Midland Bank building in Rochester in which each individual grid element up to the 6th floor is shaped so as to define and express the structural strength to take the flow of forces. The result is an expression akin to traditional bearing wall arches.

In steel buildings, the column-diagonal frame provides the most rigid tube, and this type of building acts most nearly like a cantilever sticking out of the ground as it is loaded by wind. But what if the owner doesn't want diagonals in the exterior wall? This was the problem that SOM faced when it was decided that the Sears headquarters would take the shape of a tower structure rather than a 42-story, but larger-plan building (130,000 sq ft per floor). After this, a two-building scheme was also considered—one 60 stories high, and the other 40 stories. In any event, Sears management wanted on the order of 50,000 sq

ft per floor for their own use, but smaller floor areas were felt desirable for rental tenant spaces. The final choice—as is well known—was a building of nine bays, 75 by 75 ft, or a building 225 by 225 ft at ground level. Beyond the first 50 stories (which Sears is taking) the building peaks in sets of bays, with two bays rising the last 20 stories to the F.A.A. limit of 1,450 ft at 110 stories. Total gross area is 4.4 million sq ft.

Achieving efficient frames in ultra-high buildings without using stiffening diagonals has led to the bundled tube concept, with great planning flexibility
SOM's design partner for Sears, Bruce Graham wanted to create an open, pleasant space for the plaza level which implied a tall building rather than a squat one that would take the whole site. Engineer Fazlur Khan was sympathetic to the "environment" idea, but also wanted to achieve a tall building at lower-building costs. And

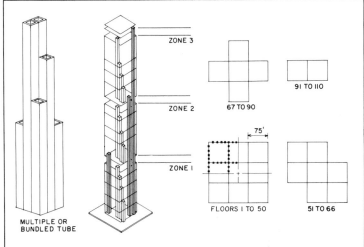

Perhaps the most intriguing concept to evolve in the ultra-high skyscraper — from both architectural and engineering aspects—is the one known as the "bundled-tube" approach, which was conceived for use in the 110-story Sears Tower. The building consists of a series of framed tubes, each of which has its own structural integrity, allowing the tubes to be dropped off as the building rises, yielding a variety of spaces for tenant floors which occur above the 50th floor. The tubes are 75-ft square, so the building is 225 by 225 ft at the base. Columns are optimally spaced 15 apart. At each corner of the tubes is a larger column that "terminates" the tube structurally with respect to wind shear transfer. Shear lag is greatly reduced, compared with an ordinary framed tube, as illustrated at right. The elevator system is divided into three zones, with two-story sky lobbies serving the double-deck elevators from the two lower zones. Sky lobbies also are served by express banks.

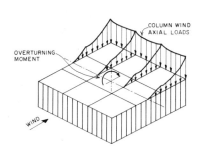

Robert E. Fischer

Graham was looking for a structural system that would let him drop off floor areas, so that part of the building would continue to rise in a prismatic way, but not the whole floor area.

With the shear-lag problem in mind, Khan conceived of putting two cross-stiffener frames (diaphragms) in each direction that would divide the building into nine cells. Then, as the building soared, cells could be dropped off, with others remaining independent. Cell size was one question. But a more important one, structurally, was that of column spacing. As the spacing gets very close (8-, 6-, 4-ft) the cost of steel and fabrications goes way up. But if columns are spaced more than 15 ft apart, the frame no longer works as a tube. So a spacing had to be found in which the cost was least, but tube action would still exist. By many parametric studies (a number of simple equations and studies) it was found that 15-ft spacing worked well, while at the same time being in accord with the building module. Computer studies showed that shear lag was greatly reduced, and that there was very little premium in square-foot costs for height. Further, there was no need to use an extremely high-strength steel (50,000 psi was highest).

With the Sears type of structure, which has been called the "bundled-tube," shear lag occurs, but it takes place in segments, which has the effect of squashing the peaks of direct stresses in the columns. What happens is that, as far as shear lag is concerned, each of the tubes appears to act independently, and the shear lag diagram drapes (like a transmission line does) from the peak at the corners, to lesser and lesser heights to the center of the building.

Because the individual tubes are independently strong with respect to wind load, they can be bundled in any sort of configuration and dropped off at will, as the building rises higher. They could be bundled five in a row and still be efficient; or placed with four around a central tube (cruciform); or have two tubes by four tubes (an L-shape). With the tube concept there is a new vocabulary of architectural space possibilities.

SOM found that concrete tube-in-tube systems, while efficient in terms of materials, were diminished in a practical sense because of the time involved to produce poured-in-place construction.

They had to find a system that has the advantages of a concrete building, but not the disadvantages. One way to eliminate the disadvantage was to make the inside of the building steel, and only the outside (lateral-stability) portion a concrete grid. What has happened is that the framed tube concept has been combined with the traditional steel frame. So far the concrete exterior frames have been made using traditional formwork as well as with precast concrete forms that were left in place to form the finished exterior. Cost savings have been $1 to $1.50 per sq ft (1962) over all-concrete buildings.

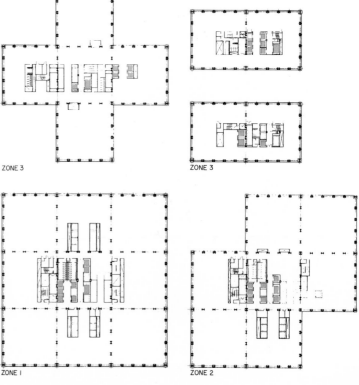

The different types of floor plans that result from "dropping off" of bundled tubes are shown below. In each of the zones, except for the top, are clear-span spaces, 75 by 75 ft. The curtain-wall system expresses the tubular nature, but not the framing of each of the tubes. While the tubes have been bundled in this particular configuration for Sears Tower, many others are possible, depending upon planning requirements. The ultimate structure, for structural efficiency, would appear to be a bundled tube with diagonals in the walls for increased stiffness.

ZONE 3

ZONE 3

ZONE 1

ZONE 2

Mast columns for Citicorp Center

A skyscraper should be a "proud and soaring thing," Louis Sullivan said, and Citicorp Center, with its lofty bearing and smooth skin, promises to be just that. As for soaring —at 914 ft, the square tower now takes seventh place among the world's tallest buildings. Architect Hugh Stubbins has, further, incorporated amenities traditional to the genre in the 1970's: landscaped plaza, shopping galleria, and a network of covered public walks.

From the pedestrian's view, the most commanding aspects of the building are its over-hanging corners, projecting 72 ft from the central columns, nine floors up. The unexpected location of the four supporting columns was dictated by the insistence of St. Peter's Lutheran Church, which shares the site, that its new building be freestanding. The church, widely known in New York as the "jazz church" because of the number of musicians in its congregation and because of its active cultural program, had occupied this corner of Lexington Avenue since 1905. St. Peter's agreement with the bank holding company in their joint development of the site was that it retain a distinct identity. On the Third Avenue end of the complex, the tower overhangs a low-rise building that houses offices and a three-story shopping galleria. On Lexington Avenue, a sunken plaza gives access to the subway and to the church's sanctuary (the granite-covered structure at street level is a large lantern above the sanctuary).

The 160-ft crown of the tower slopes toward the south in anticipation of collecting solar heat. A large solar-energy project, which was to have been funded by the Federal Energy Research and Development Administration, was abandoned after the building was designed when cost-savings proved less than hoped for. The crown, however, houses a tuned mass damper (TMD), a new and so-far unique device to slow the motion of the building in wind and so to reduce occupants' discomfort (see page 14).

CITICORP CENTER and ST. PETER'S LUTHERAN CHURCH, New York City. Architects: *Hugh Stubbins and Associates—Hugh Stubbins (principal-in-charge), W. Easley Hamner (project architect); Emery Roth and Sons.* Engineers: *LeMessurier Associates/SCI (structural)—William LeMessurier (principal-in-charge), Kenneth B. Wiesner (project engineer, and tuned mass damper), Stanley H. Goldstein (partner, New York), Joel Weinstein (design engineer, Citicorp), Fraser Sinclair (design engineer, St. Peter's); The Office of James Ruderman (structural)—Murray Shapiro (principal-in-charge); Joseph R. Loring & Associates (mechanical/electrical).* Contractor: *HRH Construction Co.*

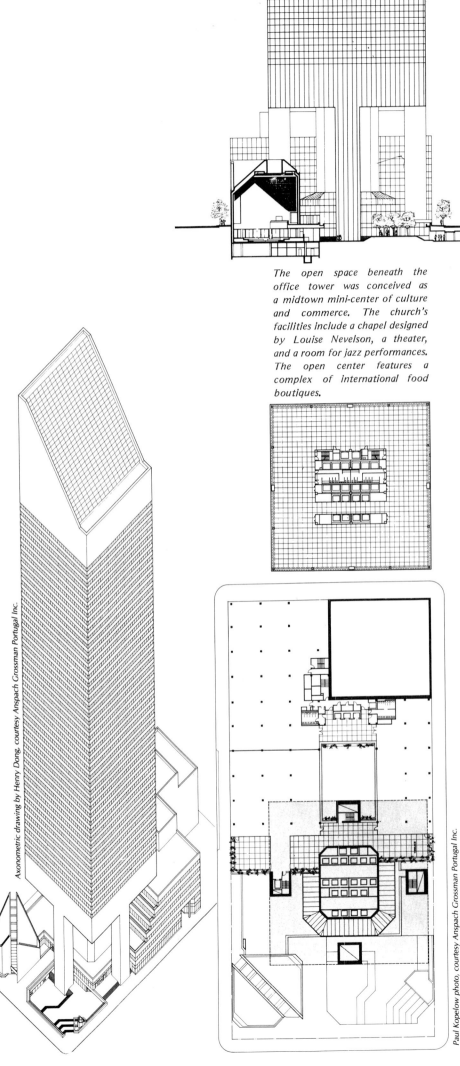

The open space beneath the office tower was conceived as a midtown mini-center of culture and commerce. The church's facilities include a chapel designed by Louise Nevelson, a theater, and a room for jazz performances. The open center features a complex of international food boutiques.

Axonometric drawing by Henry Dong, courtesy Anspach Grossman Portugal Inc.

Paul Kopelow photo, courtesy Anspach Grossman Portugal Inc.

11

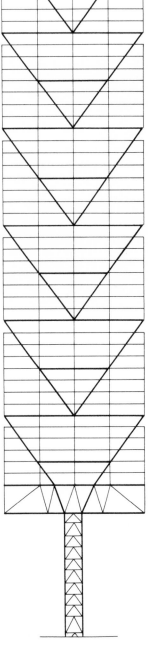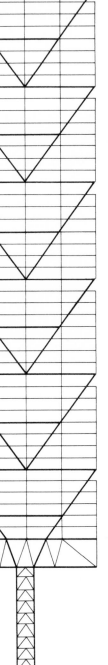

Robert E. Fischer photos

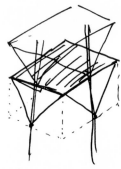

EACH EIGHT-STORY TIER IS
STRUCTURALLY INDEPENDENT...

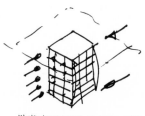

...WIND IS TAKEN BY THE CORE
FOR EIGHT FLOORS, THEN
TRANSFERRED TO THE
TRUSSED FRAME...

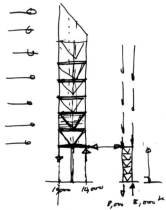

...WHICH TRANSMITS ALL WIND
LOAD TO BASE OF TOWER,
WHERE SHEAR IS TRANSFERRED
TO THE CORE.

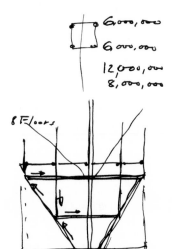

6,000,000
6,000,000
12,000,000
8,000,000

8 Floors

GRAVITY LOAD WORKS ITS
WAY DOWN THE MAST COLUMNS.

Upon reaching the bottom of the tower, the mast bifurcates at the "keystone" below top chord of the truss and transfers gravity load to the two outside columns of the legs. The 26¾-ft-deep truss, in addition to supporting the slabs for load transfer, played a major role in the construction process by providing a "getting started" platform: because the truss was constructed much as a cantilever bridge is, erection required no shoring. Below the truss, the core absorbs all horizontal shear load with four exterior bents and two interior bents (see framing plans, opposite).

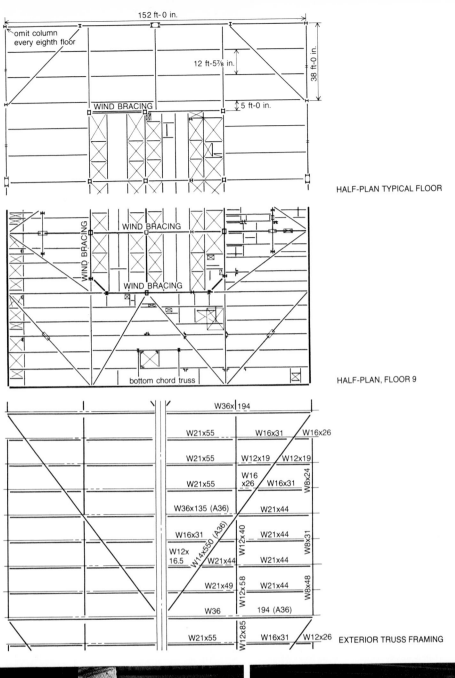

152 ft-0 in.

omit column every eighth floor

12 ft-5⅞ in.

38 ft-0 in.

WIND BRACING

5 ft-0 in.

HALF-PLAN TYPICAL FLOOR

WIND BRACING

WIND BRACING

WIND BRACING

bottom chord truss

HALF-PLAN, FLOOR 9

W36x194

W21x55   W16x31   W16x26

W21x55   W12x19   W12x19

W21x55   W16x26   W16x31   W8x24

W36x135 (A36)   W21x44

W16x31   W12x40   W21x44   W8x31

W12x16.5   W14x550 (A36)   W21x44   W21x44

W21x49   W12x58   W21x44   W8x48

W36   194 (A36)

W21x55   W12x85   W16x31   W12x26

EXTERIOR TRUSS FRAMING

Structural behavior: Even an untutored sidewalk superintendent examining Citicorp Center's unsheathed steel frame perceives that this is something new and different, something exceptional in the way of skyscraper structure. To the structural engineers, the structure represents a clean design "so simple it can be analyzed by hand," and the frame, however curious its initial appearance, does possess a straightforward and efficient elegance.

In early designs for the building, four corner columns supported the tower. This solution would inevitably have made a more or less integrated whole of church and bank and was thus unacceptable to the church (whose design contract is separate from the bank's). In a daring move, Stubbins and LeMessurier brought the columns to the center of the building face, clearing a space under the tower corner for the church.

More daringly still, most of the building's load—half the gravity and all the wind load—is brought down the trussed frame on the outside of the tower. (The remaining gravity load is carried by the core.)

On first sight, the most conspicuous members are the massive central "mast columns" and the spreading diagonals—and, on second glance, the unexpectedly slender corner columns. Only the 60-in.-wide mast column transports load down the full height of the tower. All other members of the exterior frame work only over the eight-story tier defined by the steel chevrons that feed load into the mast column.

Because the mast column accepts overturning forces, it was essential to put as much mass as possible, as quickly as possible, into this column to overcome tension. Floor load, therefore, is channeled into the intermediary columns at each level by diagonal corner beams (see framing plan), and thence to the mast column at every fourth floor. (It is the diagonal floor beams that allow the corner columns to be so slim, since they must support only a small area of the floor slab. As an almost nonchalant tour de force, the corner columns are, moreover, omitted entirely at every eighth floor, where the load is taken directly by the main structure.)

The mast column does not accept shear forces, which snake down the frame via the diagonals and ties. Within each structurally independent eight-floor tier, shear forces are absorbed by the core—a relatively inefficient necessity, although the loads are inconsequential over the short distance. At every eighth floor, these forces are gathered—"like a mother hen and her chickens," in LeMessurier's description—and taken to the exterior. At the bottom of the tower, all horizontal shear load is transferred to the core, while the "legs" carry gravity and overturning forces to the ground.

The legs comprise four columns, of which the outer two are considerably heavier than those nearest the core. According to the engineers, the legs, which measure about 17½ ft across, might have been as small as 5 ft across. They were enlarged, however, both for esthetic reasons and to house stairs, mechanical ducts and, on one side, an elevator that serves church offices and recreational spaces.

Wind damping: To slow building movement in the wind and to prevent tenant discomfort occasioned by the acceleration of the tower's sway, Citicorp will be equipped with a tuned mass damper (TMD). The device, the first of its kind in a tall building, operates somewhat as a hydraulic door closer, though it does not damp building motion simply by passive reaction but rather by countervailing movement of its own—what the engineers describe as "the application of a controlled force to a moving mass." The principle of the device is to place a large mass at the top of the building, to leave the mass "free" to remain still as the building moves, to transmit this tendency to remain stationary to the building through connections to the structure, and, further, to tune the machinery so that the period of the mass's movement equals the period of the building's movement.

The machinery for the TMD was developed by MTS Systems Corp., which manufactures machinery for earthquake simulation and for shock-testing army tanks. The TMD comprises 1) the mass—a 400-ton concrete inertia block mounted on oil bearings (a film of oil on a steel plate); 2) pneumatic springs, in which pistons act against compressed nitrogen; and 3) the control actuator (dashpot), in which energy is absorbed by oil. Because the building's natural period of movement could not be determined with precision before completion of the tower, and because conditions may change over the life of the building or with different winds, both spring and damper can be fine tuned. The spring is tuned by bleeding or adding nitrogen; this process was undertaken initially to bring the TMD into proper working order, and hereafter only occasionally. The dashpot, on the other hand, is tuned continuously when the TMD is in operation. The TMD starts up in response to a signal that the building is moving. Oil is pumped to 12 oil bearings which lift the mass block and at the same time provide a low-friction surface.

Failsafe measures include a system of curbs and snubbers which ensure that movement remains within design limits; if the snubber is engaged, the controls shut off pressure for the bearings, and the mass comes to rest.

Analysis and wind tunnel testing of the TMD indicated a 38 per cent reduction of acceleration for the tower. The engineers figured that added mass to the structure to achieve the same effect would have cost about $5 million, against the TMD's $1 million (1976).

Robert E. Fischer photos

Structural details: From Citicorp's interior, views will differ according to which floor of the eight-story tier the viewer occupies. On the top floor, corner columns (A) are eliminated where diagonals meet, while columns on remaining seven floors are unusually slender (C). On the first level, diagonals join the mast column at the center of the building face (B). The central mast is built up of rolled sections and plates (details upper right); since the depth of the web in mast column components and in diagonal members is the same regardless of the thickness of the component sections, "knuckles" match dimensions of diagonals for welding. At the fourth level, the intermediary column intersects the diagonal (D). Like the topmost corner column, the intermediary column could in theory have been eliminated at this point, but it was in fact required to resist buckling of the diagonal. This column is not, however, designed for dead load, which at this level is taken by the midpoint tie. Connection between column and tie could not therefore be bolted (detail far right) until the tier was completed and diagonal loads would not feed into the column. After connection is bolted, column D carries only live (people) load. A panoramic view of the fourth level (E) emphasizes the expanse of the column-free office floors: 46 ft from core to exterior, roughly 36-38 ft between exterior columns. The building's curtain wall is a smooth skin of pale, natural-colored aluminum banded by reflective glass. The simple shape and cool texture of the tower provides a counterpoint to the complex polyhedron, that constitutes the lantern above the church.

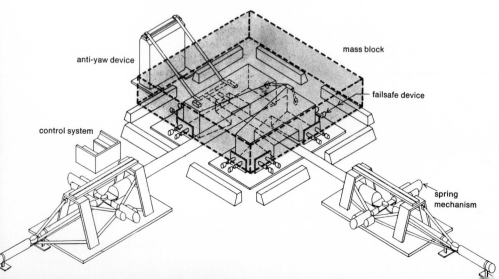

anti-yaw device

control system

mass block

failsafe device

spring mechanism

Exposed wind-brace trusses for St. Louis tower

38 ft | 30 ft | 30 ft | 38 ft

W16 | W14 | W16
W12
W18
W16
W16 | W18
W18 | W24
W16 | W14 | W16
W12 W12
W24

Stub-girder framing uses high-strength girder to span 38 ft from core to perimeter. Stub pieces—3-ft lengths of wide-flange section shop-welded to girder's top flange—alternate with continuous beams; ductwork fits between. At corners, beams extend diagonally to frame into horizontal truss members and to brace diagonal members at one of the two intermediate levels (geometry forbids this at third level). Corners are supported vertically by slender square tube sections.

core

1

STUB GIRDER SYSTEM

core

2

W/beams

CONVENTIONAL SYSTEM

10 ft

3-6 in

duct

W14 stub girder

ceiling

Section 1

10 ft

4 ft

duct

ceiling

Section 2

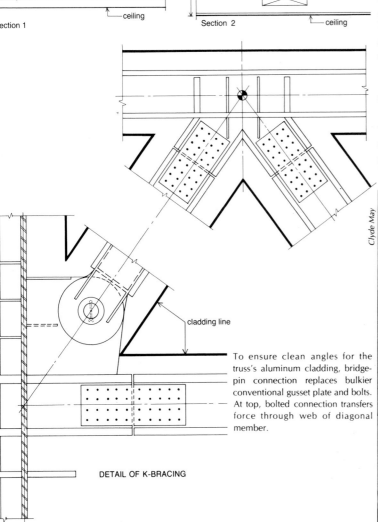

cladding line

To ensure clean angles for the truss's aluminum cladding, bridge-pin connection replaces bulkier conventional gusset plate and bolts. At top, bolted connection transfers force through web of diagonal member.

DETAIL OF K-BRACING

The monumental K-bracing at the corners of St. Louis' Mercantile Tower evolved from an interaction of architectural concern for site plan and structural concern for wind bracing of the narrow 35-story building

So that sunlight can penetrate the rather dense block of buildings, the architects cropped the corners to produce an elongated octagonal plan. From the inception of the design, the engineers worked closely with the architects to devise an exterior truss bracing system, which provides a number of structural advantages. In the system that evolved, vertical trusses were located on all four diagonal corners. This position, which according to engineer Joseph P. Colaco enhances torsional rigidity, led to the "channel bracing" at each end of the building. This "channel" is a five-sided rigid shape at the outside wall (see bold lines on structural plan), formed by welds at both ends of the short face and another weld at the bay adjacent to the trusses on the broad face. The building thus acts as a partial tube, with wind loads across the building taken by all four trusses. There is no interior core bracing, and the floor framing is designed only for gravity load; it also works as a diaphragm to transfer wind load.

Each segment of the truss is three stories high, a condition that demanded careful architectural and structural detailing. The corners of the saw-tooth floors frame into the horizontal truss members at every third level (see framing plan, top left). At the intermediate level directly below the horizontal members, the diagonals are braced in order to reduce their slenderness ratio.

In addition, connection details at the junction of diagonals and both vertical and horizontal members had to be carefully worked out to minimize bulk and simplify construction (below left).

The structural design effected considerable savings by utilizing a stub-girder system for floor framing (see drawing above left). The engineers estimate 25 per cent reduction in structural steel requirements for a conventional system, 15 per cent in structural cost. Because the depth of the floor is 6 in. less than that of conventional beams and slab, thus reducing the height of the tower some 17 ft, further savings occurred in the curtain wall and in vertical risers.

MERCANTILE TOWER, St. Louis. Owner: *Mercantile Center Associates a Joint Venture (Mercantile Trust Co., Crow, Pope & Land Enterprises).* Architects: *Sverdrup & Parcel and Associates, Inc.; Thompson Ventulett & Stainback, Associate Architects.* Engineers: *Ellisor Engineers, Inc.* (structural); *Chenault & Brady, Inc.* (mechanical/electrical). Contractors: *Mercantile Trust Construction Joint Venture (Henry C. Beck Co. and Millstone Construction Co., Inc.)* (general).

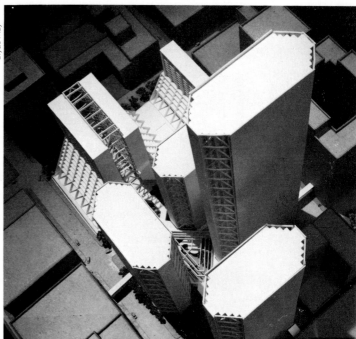

Optimized structural design approaches

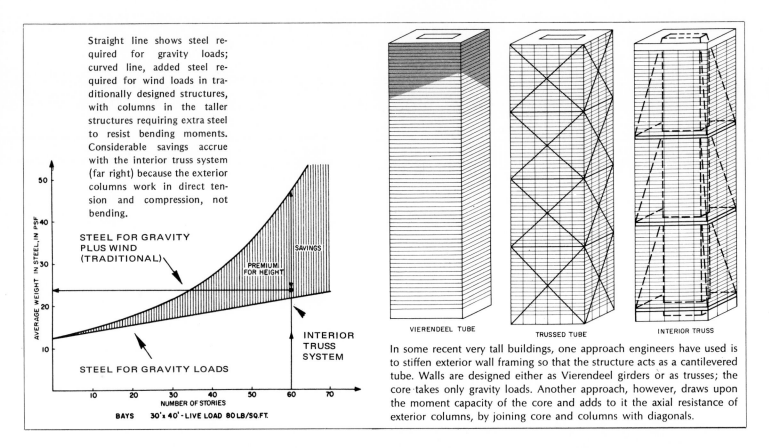

Straight line shows steel required for gravity loads; curved line, added steel required for wind loads in traditionally designed structures, with columns in the taller structures requiring extra steel to resist bending moments. Considerable savings accrue with the interior truss system (far right) because the exterior columns work in direct tension and compression, not bending.

AVERAGE WEIGHT IN STEEL, IN PSF

STEEL FOR GRAVITY PLUS WIND (TRADITIONAL)

SAVINGS

PREMIUM FOR HEIGHT

INTERIOR TRUSS SYSTEM

STEEL FOR GRAVITY LOADS

NUMBER OF STORIES

BAYS   30'x 40'-LIVE LOAD 80 LB/SQ.FT.

VIERENDEEL TUBE   TRUSSED TUBE   INTERIOR TRUSS

In some recent very tall buildings, one approach engineers have used is to stiffen exterior wall framing so that the structure acts as a cantilevered tube. Walls are designed either as Vierendeel girders or as trusses; the core takes only gravity loads. Another approach, however, draws upon the moment capacity of the core and adds to it the axial resistance of exterior columns, by joining core and columns with diagonals.

The structural design philosophy which the consulting engineering firm of Severud, Perrone, Sturm, Conlin, Bandel has evolved for the higher of the high-rise buildings could be summed up in this statement: Find ways to make all of the various structural elements work to their maximum capacity, consistent with practical construction methods; further, to utilize non-structural elements (such as mass) to help reduce movement caused by wind.

Savings in steel tonnage—and cost—can be dramatic in tall, high-rise buildings (40-50 stories and over) if certain design techniques are employed to utilize the full capacities of the structural elements. For example, with conventional wind-bracing techniques, the amount of steel required to keep drift (sway) within tolerable limits can be more than that required to withstand gravity loads. But new approaches being used by the Severud organization enable them to cut the steel tonnage back to a

little more than required just for gravity loads. The savings in steel tonnage can amount to several millions of dollars in a 60-story building.

A second aspect to be considered is that the sheer size of these buildings means any savings that can be achieved in floor-supporting elements such as beams and girders can, in total, produce significant cost reductions. With plastic design, weight reductions for beams and girders can be on the order of 35 per cent or more. Of course, connections that provide structural continuity are more expensive than those for simply-supported members. However, the savings can be very substantial, particularly if some ingenuity is applied to the design of moment connections.

New approaches to wind bracing necessary to make high high-rise buildings practical

Conventionally, tall buildings have been braced against wind by providing trussed

bracing at the core or around stairwells, or concrete shear walls at these locations, or at other convenient plan locations such as end walls, or party walls in apartment buildings.

But when buildings are higher than 500 ft or so, the core, if kept to a size consistent with elevatoring and mechanical requirements, does not have sufficient stiffness to keep drift caused by wind down to reasonable limits. Adding to the problem is the fact that these tall buildings are also slender, at least in one direction. While the buildings themselves might be stable enough, the lateral movement might be so large as to cause cracking of partitions and windows, and even perhaps to cause unpleasant psychological reactions amongst the building occupants.

If the core is not sufficient for wind load, then in long-span structures, such as present-day office buildings, the designer has to turn to the exterior framing to get stiffness. The approach engineers used in the past to achieve this was to design columns

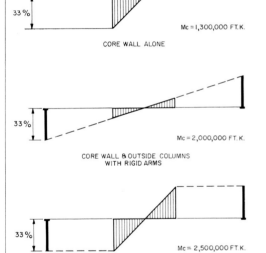

33%

Mc = 1,300,000 FT.K.

CORE WALL ALONE

33%

Mc = 2,000,000 FT.K.

CORE WALL & OUTSIDE COLUMNS
WITH RIGID ARMS

33%

Mc = 2,500,000 FT.K.

CORE WALL & OUTSIDE COLUMNS
WITH ELASTIC ARMS

WIND STRESSES

Photos show how the structure works in which a moment-resisting core is hinge-connected to exterior columns. The efficacy of the system is demonstrated in the upper-left photo in which both models have the same lateral load applied. The unrestrained core deflects much more than the "tied" core (core is of sponge rubber). Note that there is no bending in the columns in the tied-core system. The columns and truss keep the core from rotating at the top, causing it to bend as shown. The photo above right shows the behavior of such a structure with two ties. The drawing at left shows how the moment capacity of the core and engaged columns is nearly twice that of the core alone. The assumption made in the drawings is that the capacity available in the columns for wind load is 33 per cent of that for gravity load. It can be seen in the middle diagram that when the exterior columns are rigidly connected to the core, the usable moment-resisting capacity of the core is reduced considerably from the situation where the core is hinge-connected to the columns.

and girders as rigid frames (wind bents). But while the building had the desired stiffness, the structural design was disadvantageous in that big bending moments were thrown into columns, particularly in office buildings which required large column-free areas between the outside wall and the core. Thus the columns became huge and steel tonnage could reach an amount as high as 60-65 lb per sq ft.

More recently, structural engineers conceived the idea of designing very tall structures so that they behave like hollow-tube cantilevers in resisting wind forces. The structural framing is made stiff in the plane of the walls to take wind loads, while the core is designed only for gravity load. Wind blowing perpendicular to one side of the building is transmitted via stiff floors acting as diaphragms to the two walls parallel to the wind which resist the load in shear. In buildings finished to date, the exterior walls have been made stiff by one of two methods: 1) designing the exterior framing as a Vierendeel girder, or 2) designing the exterior wall as a truss with diagonal bracing. The advantage of the first is that the exterior wall is rectilinear, and more familiar esthetically; its disadvantage is that it takes more steel than the second approach because local bending is induced in the members. The big advantage of the second approach is its structural efficiency with wind forces being resolved mainly by tensile and compressive forces; the disadvantage, if it is so to be considered, is that the architect of necessity must work with diagonal elements in the design of his exterior wall.

Recently a third approach has emerged. Severud has employed it in the 51 story I.D.S. Center tower in Minneapolis, designed by architects Philip Johnson and John Burgee. This approach draws on the inherent, but in other approaches sometimes latent, strengths of both the core and the exterior framing, but it circumvents inefficiencies inherent with wind bents.

Severud partner Hannskarl Bandel points out that inasmuch as codes allow stresses in columns to be increased by 33 per cent over those permitted for gravity loads to accommodate wind load, the design approach should use this reserve capacity as efficiently as possible. Furthermore, he says, the core also should be made to work as a wind-resisting element.

The approach, as applied in the I.D.S.

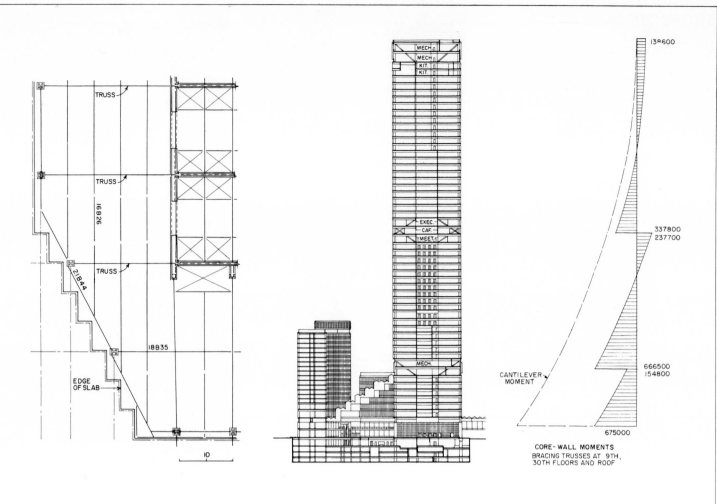

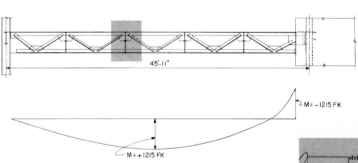

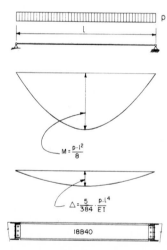

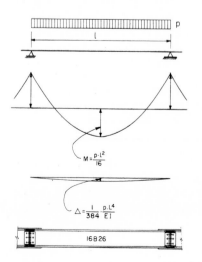

TRUSS

TRUSS

16B26

21B44

TRUSS

18B35

EDGE OF SLAB

10

MECH.
MECH.
KIT.
KIT.

EXEC.
CAF.
MEET.

MECH.

13B600

337800
237700

CANTILEVER MOMENT

666500
154800

675000

CORE-WALL MOMENTS
BRACING TRUSSES AT 9TH,
30TH FLOORS AND ROOF

45'-11"

M = -1215 FK

M = +1215 FK

p

l

$M = \dfrac{p \cdot l^2}{8}$

$\triangle = \dfrac{5}{384} \cdot \dfrac{p \cdot l^4}{EI}$

18B40

p

l

$M = \dfrac{p \cdot l^2}{16}$

$\triangle = \dfrac{1}{384} \cdot \dfrac{p \cdot l^4}{EI}$

16B26

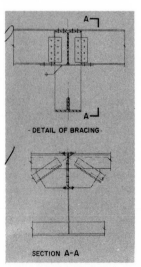

- DETAIL OF BRACING -

SECTION A-A

The 51-story I.D.S. tower in Minneapolis, designed by the architectural firm of Philip Johnson and John Burgee has a number of innovations for reducing the weight of the steel in the structure to a remarkably low 17 lb per sq ft. The first of these is the use of interior trusses, occupying two to three levels at mechanical floors, that tie the core to the exterior columns (see drawing above, center). As shown on the previous two pages, this engagement of core and exterior columns allows these columns to work in direct tension and compression rather than bending—providing a considerable increase in the efficiency of the steel frames.

Another technique that saves steel is the use of moment connections at the core of the building for the truss girders that are the main framing for each floor. They are simply supported at exterior columns so as not to induce bending in these columns. This support method reduces bending moment caused by gravity load (drawing above). Filler beams spanning between the girders have continuous connections to reduce bending moment (see drawings far right). A suggested connection for this continuity is shown in the top detail in the gray tone, right.

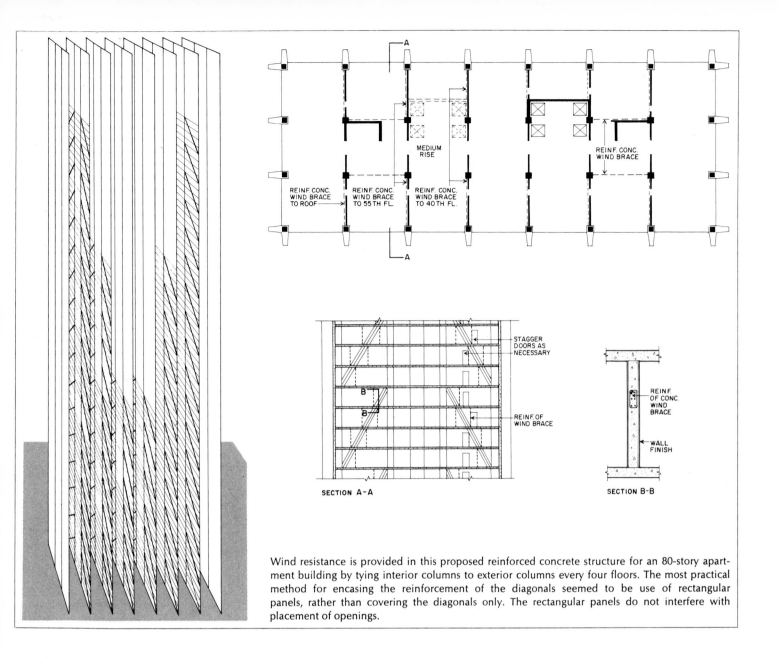

A

MEDIUM
RISE

REINF. CONC.
WIND BRACE
TO ROOF

REINF. CONC.
WIND BRACE
TO 55TH FL.

REINF. CONC.
WIND BRACE
TO 40TH FL.

REINF. CONC.
WIND BRACE

A

STAGGER
DOORS AS
NECESSARY

REINF. OF
WIND BRACE

B

B

REINF.
OF CONC.
WIND
BRACE

WALL
FINISH

SECTION A-A

SECTION B-B

Wind resistance is provided in this proposed reinforced concrete structure for an 80-story apartment building by tying interior columns to exterior columns every four floors. The most practical method for encasing the reinforcement of the diagonals seemed to be use of rectangular panels, rather than covering the diagonals only. The rectangular panels do not interfere with placement of openings.

building, and in others the Severud firm has under design, is basically this:

The core is made a moment-resisting element by being designed as trussed framing, concrete shear walls, or a composite of the two. The core is connected to exterior columns at two or three levels in the building by means of diagonal trussing (usually where there are mechnical floors, computer rooms, kitchens, etc., where diagonals will not interfere with space usage). These trusses are connected "elastically" to exterior columns; in other words the connections effectively are hinges. The reason for this is to have the core engage the exterior columns, but not throw any bending moment into them when the structure is loaded by wind. The behavior of the system is best illustrated by looking at the drawings and diagrams on the first two pages.

The core by itself (assuming no surrounding structure) if loaded laterally would deform as a free cantilever. But when the core engages the exterior columns it is no longer free to move in this manner. The trussed framing tries to rotate with the core, but it cannot do so because it is restrained by the exterior columns. The columns neu-

tralize the rotational force by a compressive reaction in the set of columns on the leeward side of the building and by a tensile reaction on the windward side.

An ancillary advantage of the interconnection between interior trussing and exterior columns is that the tendency for differential elongation caused either by temperature effects or creep and shrinkage can be largely equalized between outer columns and interior cores—the diagonal bracing between vertical members does not allow free deformations.

Cutting down the cost of the structural elements supporting floors

The design of beams and girders utilizing high-strength steels and plastic theory permits considerable reduction in steel tonnage required. Elastically designed beams and girders using high-strength steel would have larger deflections than with ordinary-strength steels because of their lesser depth, and thus lesser moment of inertia. This potential problem is overcome when beams and girders are designed with continuity which cuts deflection to less than half that of simply supported members. But a big problem

in working out a plastically designed floor system is detailing the moment connections in such a way that they are not so complicated to fabricate as to wipe out the savings accomplished by the reduction in weight of steel. A possible solution for connecting plastically-designed filler beams to trussed girders is shown on page 20.

Coping with the space requirements for lateral runs of large air-conditioning ducts further complicates the problem of designing the floor system. Trussed girders provide large openings, but they are more expensive to fabricate than rolled sections.

The continuity between filler beams does not affect other structural elements in a deleterious way. But, if girders between core and exterior columns were made fully continuous, they would create bending moments in the columns, necessitating a larger size than would be required for gravity and wind loads. But, on the other hand, a rigid connection can be made at the core which has the capacity for moment connections, and the girder works as a cantilevered member simply supported at one end. This means of support has nearly 70 per cent the efficacy of a member fixed at both ends.

Wind damping approach at the World Trade Center

Doors have dampers to slow their closing. Cars have shock absorbers to smooth out the ride. But until now, buildings have not had engineered dampers, as such, to take out the energy that the wind pumps into a tall building. The building in point is the World Trade Center whose two twin towers each have 10,000 dampers at the ends of floor trusses.

Today's buildings are taller, have longer spans, and are more flexible—thus they are affected more than before by the vagaries of the wind. When wind blows on a tall building it puts energy into it that, in one way or another, must be dissipated to avoid distress to the building structure and discomfort to the occupants.

The earlier skyscrapers had heavy walls and partitions that added mass and damping to the structure. Further, their facades usually had strong texture which served to create turbulence, and thus acted as a natural means of damping. The exterior walls and partitions sliding across contiguous parts used up energy by means of friction. Of course, it is still possible to utilize textural devices. For example, the U. S. Steel Building in Pittsburgh is a triangle in plan, but it has notches cut out of the corners which create turbulence when the wind blows.

But buildings have reached such heights and sizes that occasionally natural damping may not be enough. The structural engineers for New York's World Trade Center, Skilling, Helle, Christiansen, Robertson, elected to find ways to diminish not only the swaying motion from the strongest winds, but also the more average oscillations that could make maintenance costly—cracked partitions, perhaps broken glass, etc. The simple geometrical shape of ·the two 110-story towers was not ideal from the standpoint of creating air flows that would tend to break up vibration patterns.

First, the strong architectural expression found in the chamfered corners of the tower structures modified the generation of vortex excitation thus reducing structure motions. Next, the structural engineers found that an economical and reliable method to use up the wind's energy would be to install non-structural energy absorbers at the ends of the bottom chords of the floor trusses. The idea of hydraulic dampers (similar in a way to a door closer) was considered, but these would have had

to be excessively large, and they would have been expensive. A lot of liquid would have to be moved to absorb the requisite energy.

The most promising approach seemed to be that of viscoelastic damping. As the name implies, viscoelastic materials are both viscous and elastic; they are both liquid-like and spring-like. A viscoelastic material is elastic to the extent that it will return to its original shape when deformed at moderate rates. Viscoelastic materials use up energy by resisting forces in shear. Very thin layers of these materials are very effective energy absorbers. In optimum damper design, after the viscoelastic material has been strained in one direction, the return is just slow enough to oppose the next cycle of oscillation. Most of the energy input to the viscoelastic material is not stored, but is used up, being converted to heat. A "perfect" spring, in contrast, stores all the energy and puts it right back into the object which applied the force to it in the first place.

In developing a damper design for the World Trade Center, the structural engineers worked with a technical research team from the 3M Company which proposed the use of an acrylic copolymer for the viscoelastic component of the damper.

Approximately 10,000 dampers were installed in each 110-story tower. About

100 dampers were installed at each floor, from the seventh through the 107th.

The damper is composed of three steel elements, two 4-by 4-in. tees and one ½-in. by 4-in. bar between which are sandwiched two viscoelastic layers—0.050 in. thick by 4 in. by 10 in. These layers are epoxy bonded to the steel. One end of the bar extends beyond the bonded area for attachment to the bottom chord of the truss, while the coped ends of the two tees extend in the opposite direction for bolting to a seat on the column.

When the building is "excited" by a gust of wind the dampers ordinarily move only a few thousandths of an inch, but are designed for as much as 0.02 in. movement. The dampers must dissipate at least 300 in. lb of energy per cycle at maximum deflection.

The structural engineers chose the bottom chord location for the dampers because it offered a "large" (relatively speaking) available displacement. Further, it caused no physical interference with any of the other building components. The dampers could not be too stiff, or the wind forces would have resulted in undesirable flexural stresses. On the other hand, if the dampers were too weak, the energy absorbing properties of the dampers would have been considerably reduced.

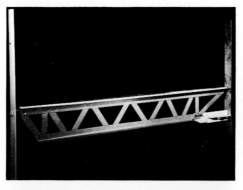

This model simulates the effect a damper has on diminishing the oscillation of a structure that has been set in motion by the force of the wind.

When the frame is undamped (top), the energy is only slowly dissipated, so the frame swings farther to the right than when it is damped (bottom).

Wind effects: the one-two punch

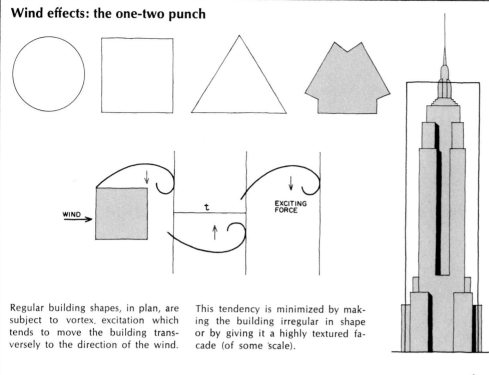

Regular building shapes, in plan, are subject to vortex. excitation which tends to move the building transversely to the direction of the wind.

This tendency is minimized by making the building irregular in shape or by giving it a highly textured facade (of some scale).

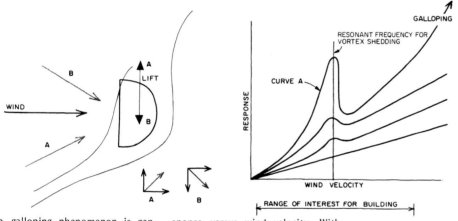

The galloping phenomenon is generally associated with transmission lines coated with ice. The wind creates both lift and drag. Because of lift, the body first sees the wind from direction "A" and then from direction "B".

The graph, right, plots building response versus wind velocity. Without any damping the plot would be like curve A. With damping added, the effect of vortex shedding is decreased with each increment of damping. At a certain wind velocity (generally not a condition experienced by buildings) galloping starts.

The elastic restoring force of a strained viscoelastic material results from the distortion of chemical bonds and the decrease in the number of possible conformations due to uncoiling of polymer chains. The viscous resistance to deformation is due to the liquid-like drag which the chain segments encounter as they move relative to each other. The type of viscoelastic material used in the World Trade Center dampers lends itself best to constrained layer or interlayer damping because it has more damping per unit volume than other damping materials, and because of its bonding capabilities, either through inherent tackiness or through ease of bonding with epoxy adhesives. The materials are somewhat temperature-sensitive in terms of efficiency of energy dissipation. Exhaustive tests were run on sample dampers and on production dampers at the 3M Company and at the Massachusetts Institute of Technology to find out what happened to the heat. Actually only a minute temperature rise (measured using thermocouples) took place in the material as it was strained in the testing machine. The heat was quickly dissipated by the steel plate and the tees. Further, in actual use, the dampers will be in a controlled temperature environment within the building—they won't get the hot summer sun, nor the cold winter wind.

For measurement of the physical performance of the dampers—i.e., their efficacy in dissipating energy—a graphical plotter was used to draw the graph of load versus extension. The plot is in the form of a hysteresis loop, the area of the loop being a measure of the energy dissipated (see picture of actual graph page 25).

Four kinds of damping attenuate the wind's energy in buildings

There are four major kinds of damping associated with building structures: 1) material damping from the structural frame, 2) damping from partitions, exterior walls,

Approximately 10,000 dampers were installed in each 110-story tower of the World Trade Center. A damping unit is comprised of two tees and a plate between which are bonded two layers of 0.05-in. viscoelastic material. One end of the bar is attached to the bottom chord of the truss; the ends of the tees are bolted to a seat fastened to the column. Dampers were custom designed and fabricated to the structural engineers specifications.

etc., 3) aerodynamic damping, and 4) added damping.

Material damping intrinsic in the structure itself is usually very small; perhaps a fraction of one per cent of critical damping because structures, basically, are very elastic. (See drawing, page 25).

Damping found in non-structural elements results from sliding of materials, one past the other, as buildings move. Examples are: the movement of the glass wall in its rubber edge; the friction of partitions as they slide; the cracking of non-structural parts, such as masonry shaft enclosures.

Aerodynamic damping is very much associated with the shape of the building. Irregularly shaped buildings have higher aerodynamic damping than smooth, uniform shapes. Some have negative aerodynamic damping; that is, they not only do not damp oscillations, they enhance the tendency to oscillate.

Excitation results from various phenomena, one of these being vortex shedding. Excitation also is manifested in the galloping phenomenon which commonly occurs with conductor lines. Because of drag and lift, oscillation is increased.

A wind force on a body (particularly one of oblong shape) moving in a given direction may produce a lift force, just as a lift force is created on an airplane wing or a boat sail. If the wind tends to push the body in the direction it is already moving, then the body travels farther and faster. And if the lift force further increases, the body tends to move still faster. Then if the process is reversed it will oscillate, or gallop, as high-voltage lines do.

Vortex excitation is different. Vortex excitation results from the shedding of vortices in a periodic relationship—with the shedding frequency being dependent on the shape and size of the building. If this frequency happens to be near the natural frequency of the structure, then it will oscillate back and forth transverse to the wind flow.

The fourth kind of damping is induced or added damping. This is utilized sometimes in aircraft and other kinds of structures. It is used all the time for door closers and automobile shock absorbers. These are basically added damping systems. You could let the door slam, relying on the intrinsic damping that comes from the hinge and various other things, or you could add something in. This is what was done in the World Trade Center.

The damping system involves a structure which is separate from the basic structure of the building. All the dampers could be taken out and the building would stand there; people could live in it and work in it without difficulty. But without dampers, some people would perceive building motion during periods of high wind.

High buildings need damping so the wind won't bother structures or people

Why damping in the first place? The reason is, if you make a plot of the dynamic component of structure response (not the steady state component, which damping does not influence at all) on the vertical axis against the velocity of the wind on the horizontal axis, you get a kind of monotonic increase that corresponds to velocity— generally not logarithmic but monotonic (see page 25). And the slope of this curve is related to the amount of damping. The more damping, the flatter the curve. The less damping, the higher the curve. This dynamic motion is a sum of phenomena mentioned earlier, galloping, vortex shedding, and so forth.

The picture is confused by the fact that over the height of the building, the wind velocity changes. So, the excitation comes at many frequencies, at many levels. And, further, if the building is irregular in shape, you get an additional complication of these exciting frequencies.

There is not any single velocity that a building sees. It's a jumble of excitations—

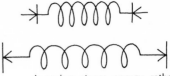
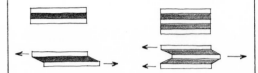

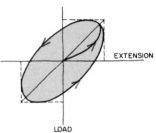
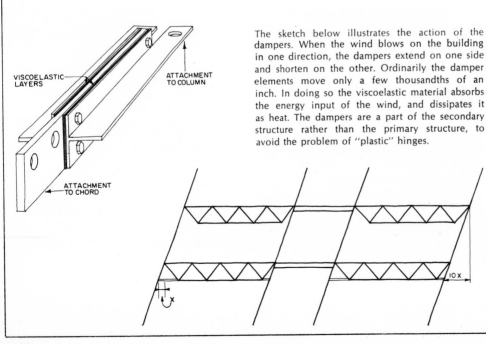

VISCOELASTIC LAYERS

ATTACHMENT TO COLUMN

ATTACHMENT TO CHORD

The sketch below illustrates the action of the dampers. When the wind blows on the building in one direction, the dampers extend on one side and shorten on the other. Ordinarily the damper elements move only a few thousandths of an inch. In doing so the viscoelastic material absorbs the energy input of the wind, and dissipates it as heat. The dampers are a part of the secondary structure rather than the primary structure, to avoid the problem of "plastic" hinges.

10 x

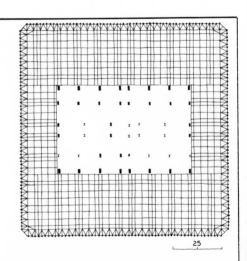

25

turbulence induced, galloping induced and vortex induced—which generally prevents the "pure" response shown in curve A of the graph on page 23.

There is a critical wind velocity which, blowing uniformly, would produce the greatest dynamic excitation and tend to produce a peak response. Damping tends to level that peak out. Damping of all kinds tends to level it out. Aerodynamic damping, particularly, tends to make it very broad because of the fact that aerodynamic damping comes also where the velocities are highest.

If a building does not have enough damping, a high dynamic response may result. High dynamic response can mean several problems; it can mean the total response (deflection, acceleration, stress, etc.) is high. If the total response is high, the building must be designed for a greater excursion and, consequently, higher stresses in all the pieces. Further, all the architectural finishes (cladding, partitions, etc.) have to be detailed to accommodate the greater extreme. So, one aspect of damping is that excursion is reduced.

The other is that the perception of people to that motion is reduced. It is interesting that for a given shape of building, given size, and given wind environment, there is almost nothing that will reduce acceleration except to add damping or to add mass. Increasing the stiffness reduces the excursion, but you have the same level of acceleration along with the lower excursion because the frequency goes up. And also as the frequency goes up, people are, at least marginally, more sensitive to that motion. People are more susceptible

to the higher frequency motion than the lower frequency motion. Why? Because higher frequency motion has a higher rate of change of acceleration. The rate of change of acceleration is perhaps more of a factor in perceiving motion than acceleration itself.

When the stiffness of a building is increased people become more sensitive to the acceleration. So you come back to only two things that are really available—damping and mass. But mass is very expensive. The logical vehicle the designer should search for is the kind of damping that is free. And the best kind that is around is aerodynamic damping. Considerable aerodynamic damping can be achieved by doing many things that architects like to do. For example, you can cut out the corners and get increased aerodynamic damping. Building shapes can be compared for their relative response.

Partitions add stiffness and they add damping. This is not a linear kind of stiffness however. Non-linear stiffness means damping. On the older buildings a lot of stiffness was available from partition systems as well as mass and damping.

The adding of damping, which is measurable, reliable and which, so to speak, can be designed into the structural system has obvious advantages. The ways available for adding damping are not very many.

Damping devices can be used in trusses, or girders, or beams, or almost anywhere there is relative motion between pieces.

The wind can blow and blow at a given velocity for hours—not for seconds, portions of minutes, but for hours. Energy

just keeps pumping into a building like soldiers walking across a bridge—they just keep pumping energy into it. But if the wind keeps putting energy into a building with insufficient damping, eventually the point is reached where it is oscillating too much.

The amount of energy that goes out (i.e., dissipated) has only to do with the amount of damping in a building—that is what damping is. It is the using up of energy of oscillation. If there is not enough damping, then the building sway will be out of bounds and the structure will start moving to the yield point. Then damping will come not from controlled sources but from uncontrolled sources, and fatigue, bent structures or other permanent damage will result.

Floor vibration can be damped using engineered materials

Another promising use for dampers is for floor vibration. There are three parameters to work with. First is the mass of the floor. It can be designed heavier. Second; it can be made stiffer—usually by deepening the floor to get the desired stiffness. When the floor is stiff, vibrations still exist, but they are of sufficiently low magnitude. Third, damping can be added.

If the designer is dealing with, say, a mezzanine in a store, the addition of depth could be very very costly because it would influence the whole building, not just the mezzanine. So the designer might very well find it economical to put damping into the floor. Damping doesn't prevent vibration. But, importantly, what it does is reduce its amplitude as quickly as possible.

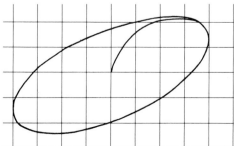

A trace of an actual "hysteresis-loop" plot made by the graphical plotter connected electrically to the testing machine shown at top.

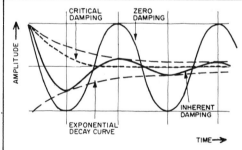

FREE OSCILLATION WITH VARIOUS DAMPING RATIOS
With no damping a body will continue to oscillate at the same amplitude. Increased damping decreases the amplitude. With critical damping, the structure stops before one cycle.

Sample of the viscoelastic material used in the World Trade Center dampers. Note how limp it is, conforming to the shape of the person's hand.

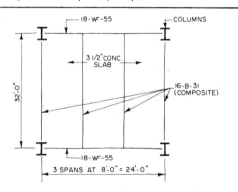

This is an example of viscoelastic material being used to control the vibration of a light structure used for a mezzanine in a store. It was applied to the bottom flange of the steel beam, and then a constraining layer of steel was put over it to make the material work more efficiently in shear. Dampers *do not prevent* vibration, but they do reduce its amplitude.

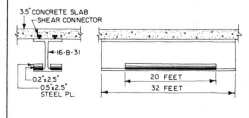

CHAPTER 2:

Structures—
Those Built with Minimal Materials

Fabric structures are now being used by architects for permanent buildings, and their application is being extended to a much broader variety of buildings than simply sports facilities, amphitheaters, and exhibition buildings—though these were natural applications whose development gave engineers opportunities to originate new design approaches and new analytical methods, and to exploit new weather-resisting, fire-resisting, high-strength fabrics.

In contrast to other types of structures, form is *everything* with fabric roofs. Inflated structures have to expand until they are taut, but their shape needs to be of such nature as to minimize stresses in fabric and cables, while also having a profile that offers little resistance to wind.

With the tent-type tensile roofs, the shapes must be mathematically based so that a) the stresses in fabric and steel cables can be determined accurately, and b) the roof fabricator can lay out his patterns so that after the poles have been erected and the edge cables tied down, the roof will be smooth and free of wrinkles.

The principles of fabric roof structures are presented in this chapter along with a number of recent, noteworthy applications.

Engineering discipline of tent structures

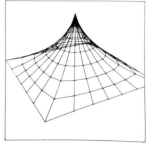

by Horst Berger, principal
Geiger Berger Associates, P.C.

The possibilities of tensile structure are tantalizing to consider: minimal use of materials and stimulating forms to provide shelter.

But there is more to their design than meets the eye. For example, the simple tent roof in the small drawing would not seem to require much mathematics to design. This is not just any garden-variety tent, however. The radial cables are not merely draped, but are prestressed a predetermined amount to give definite form to the structure and to let it withstand wind and snow loads. The engineer had to know what the stresses would be in the fabric so that they were within safe limits, and, furthermore, so that the fabricator could pattern the fabric to the right dimensions.

It follows that there is an engineering discipline that needs to be observed if materials are to work efficiently, and if shapes are to be accurately determined. If shapes are arbitrarily arrived at (either pragmatically or subjectively), then it is difficult, if not impossible, to predict stresses accurately, and the design may have to be over-engineered, or may even fail because of overstress or flutter.

Done correctly, most pole-and-frame-supported tensile structures must be analyzed structurally by mathematical methods, and to do this quickly and accurately, a computer is required. The author of this article, Horst Berger, developed a mathematical procedure, programmed for computer, that allows him to predict accurately the shapes that result from the prestress forces of self-supporting tensile roofs, and the stresses anywhere in the materials. The evolutionary process of his thinking, and of some of his designs, are the subject of this article. It is clear that the engineering discipline is no barrier to the visual imagination.

This is a computer scope picture (in perspective) of architect William Morgan's proposed Interama Amphitheater in Miami— a five-module tent. It was drawn by computer—the input being the engineer's predetermined location of points on curves in space (Cartesian coordinates) from an initial assumed shape. Pictures such as these are useful to both architect and engineer. They help the architect explore spaces— what they look like, clearances, etc. They help the engineer visualize flow of forces in complex three-dimensional shapes. The engineer knows in which direction cables of a network must pull for a shape to be stable. A computer picture may tell him that not all his initial assumptions were correct: with the assumed shape, the cable geometry may not result in forces pulling in all the right directions. With Interama, Horst Berger was easily able to modify the geometry where the end wing met the first module in order to get the proper stress flow.

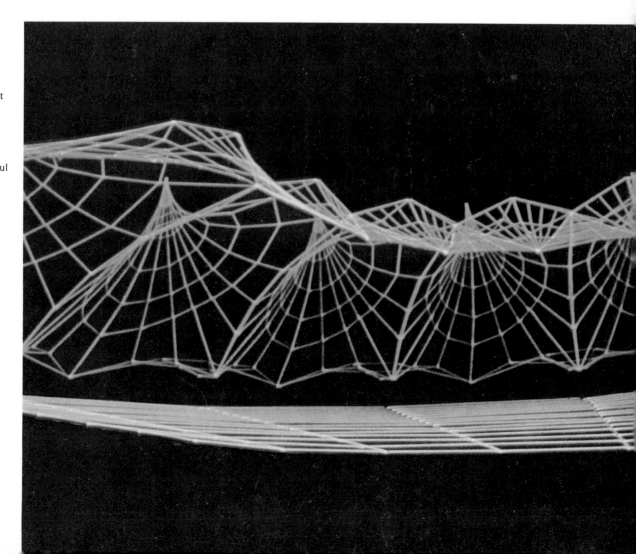

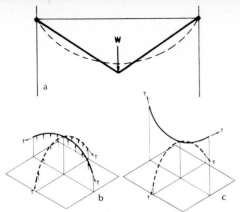

Figure 1: the force discipline of tensile structures

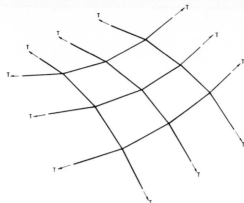

Figure 2: tent structures must be doubly curved

Figure 3: air structures are singly curved

Only in the last 25 years have suspended roof structures been used for substantial buildings. New materials such as fiberglass, plastics, and fabrics of exceptional strength and elasticity led to the now popular air structures. More advanced designs were developed for the international fairs of recent years. The U.S. Pavilion at Osaka, for example, led to permanent, low-profile air structures of fabric and cables, an advance achieved through use of a durable, fire-resistant membrane, simplified construction, and highly improved design methods. The largest of these, for Pontiac Stadium, covers 376,000 sq ft.

For large spans, air-supported buildings remain one of the most efficient applications of modern fabrics and coatings. Where height is desirable or acceptable, and where the interruption of floor space by vertical supports is not objectionable, a series of modular tent structures offers an exciting alternative.

The main obstacle to a breakthrough for self-supported tensile structures lay in the area of design—how to obtain, with ease and accuracy, the exact shape of such structures at given stress patterns and levels. The answer had to be a mathematical modeling of great adaptability which would also permit the designer to find and explore new shapes and combinations of shapes. Such a method was recently developed by the author. This article explains its background and basis.

The behavior of fabric and cables imposes unique structural conditions

The materials used to build tensile structures— fabric and high-strength cables—present two unusual structural conditions: they are extremely light, and, unlike beams, arches or trusses, they have no rigidity.

Further, unlike a beam, an arch or a truss, cables and fabrics have no stiffness. They can therefore carry load only in tension, and they must be kept in tension at all times if they are to remain stable. The stability of a tensile structure must be achieved solely by a combination of shape and prestress.

Unlike a beam, a cable changes its shape when a load is applied (Figure 1a). Any new load case will cause the cable to take on a totally new configuration. A single cable, or a tensile membrane with a single curvature, cannot, therefore, generate a stable structure. If the designer wants to avoid adding weight and stiffening members (as in suspension bridges), he must use other means to provide stability—he must shape and stress tensile networks.

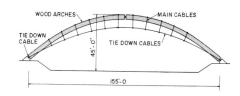

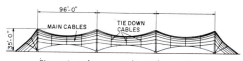

Figure 4: arch-supported tents for tennis courts

Figure 5: a frame-supported tent experiment

Figure 6: a more sophisticated tent in model form

Curvature, prestress and support system control the shape of tensile structures

The generating principles for shaping lightweight tensile structures are simple, but different conditions determine different shapes for air-supported and for self-supported structures. In an air-supported structure (Figure 1b), an external force (air pressure) acts outward and normal to the membrane surface. The sum of the normal components of the cable forces at a node must therefore act inward and be equal to air pressure. Cable nets or membrane surfaces will, therefore, curve in the same direction—that is, down toward the edges.

The self-supported structure (Figure 1c), on the other hand, has by definition no external force acting on it, because dead weight is negligible. The sum of the normal components of cable forces must therefore be zero, which

requires that two intersecting cables, or two main stress lines in a tensile membrane, be of opposite curvature. This is the first and most important principle.

If we now add a superimposed load—let us say, wind pressure—the structure must remain in tension. This requires that the cable net or membrane be under sufficient initial tension—which means it must be prestressed. This is the second principle.

Curvature and prestress must be selected to comply with this principle. The greater the curvature, the lower the prestress required. A portion of any self-supporting tensile structure therefore resembles Figure 2. Any surface configuration that permits two sets of continuous stresslines of opposite curvature is suitable for a self-supporting tensile structure. This leaves only one additional consideration: the support.

Because the two sets of cables in air-supported structures are of equal sign, it follows that they require only one set of supports, usually at the lower end of the structure (Figure 3). Self-supported structures, by contrast, require two sets of supports—one above and one below the tensile structure.

Two types of support are possible: line supports and point supports. The arrangement of the supports largely determines the configuration of the structure.

Figure 4 shows a structure supported on parallel laminated wood arches. The rigid arches form the upper supports, catenary end cables the lower supports, and an orthogonal two-way cable system is stressed between them. The fabric is attached directly to the cable system.

The prestressed fabric dome in Figure 5 is

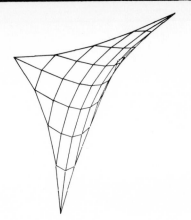

Figure 7: hyperbolic paraboloid, one basic shape

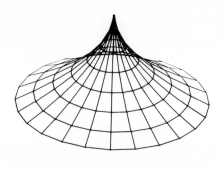

Figure 8: the radial tent, the second basic shape

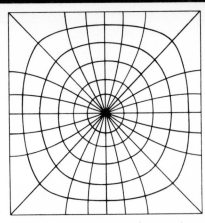

Figure 9: stress lines for a tent with a square base

a more sophisticated structure. The upper supports are still arches, here made of flexible aluminum bands which, when lifted by the center ring, take on the proper dome shape. The tensile membrane is formed by the radial valley cables in one direction and by the fabric in the other. The lower supports are the foundation points. They are interconnected with the arch supports by the triangular system made of aluminum pipes. A tension ring cable at the spring line of the arches equilibrates the arch thrusts. Thus the entire flow of forces is balanced within the system.

The prestressed fabric shelter of Figure 6 goes a step further. The compression members are completely separated from the tensile membrane, and provide four upper and four lower support points interconnected by metal triangles. The lower supports are at the end of the valley cables, and the upper supports are catenary cables spanning the distance between the high points of the triangles. (This ridge cable could, of course, be regarded as part of the tensile membrane, in which case the distinction between line and point disappears.

For a larger structure—one requiring more than a single valley cable per bay—each quadrant of the tensile membrane takes on the shape shown in Figure 7. This surface, a hyperbolic paraboloid, is one of the basic shapes that satisfies the requirements of point-supported tensile structures. Most of Frei Otto's tent structures use this surface, shaped by a net of orthogonal cables supported by diagonal edge catenaries.

The other major family of tensile structures derives its shape from one set of cables originating from the point support in a radial pattern, and a second set of cables around the point support, producing a conical shape. The simplest form has a circular base and circular rings (Figure 8); it is easily analyzed and has been used for several major structures.

The design process in reverse: shape derives from applied forces

The configuration of a radial tent with a noncircular boundary would ideally resemble the stress-flow diagram of an elastic membrane with point supports. Figure 9 shows the main stress lines in such membrane, with radial lines curving down to a square base. Because radial cables cannot reasonably be made to curve in the plane of the membrane, a practical layout is achieved by straightening the radial cables and letting the rings form polygons (Figure 10).

To build a roof structure of this type re-

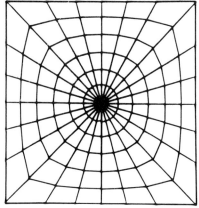

Figure 10: radial cables are straight for practicality

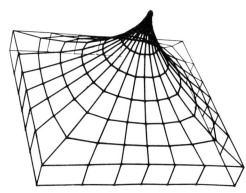

Figure 11: computer perspective of a square tent

Figure 12: a square tent for Great Adventure

quires a design tool that will determine the exact geometry of the roof, with every point of the cable system in equilibrium under a given stress. While a closed mathematical solution—one that gives finite answers in one step—is possible for circular tents, other shapes, such as the square-based tent in Figure 10, require a totally different approach. Such an approach is based on a reversal of the usual design process by predetermining the stresses in a cable-net system and letting the net find its corresponding shape. Figure 11 shows an isometric view of the tent roof designed for Great Adventure Park in New Jersey; this shape was generated by the following process:

First, the configuration of the radial cables was determined in plan. One requirement of this plan was to place the cables close enough to each other to prevent overstressing the fab-

ric spanning the space between them. Another was that the cables be numerous enough and close enough to produce the appearance of continuous curves rather than flat polygons. Still a third requirement was as much repetition in patterning as possible. A layout of 24 cables, spaced at equal angles of 15 deg, best satisfied these and several other design criteria.

Next, the number and elevation of the rings was established. Since in this structure the function of the ring cables was assumed by the fabric itself, the rings were simply a mathematical tool. As a first step towards producing a structure with fairly uniform fabric stresses, these hypothetical rings were located so that their average spacing was approximately equal. In this case, it was decided that five rings provided sufficiently close spacing for accurate stress analysis and fabric patterning.

Next, a set of compatible forces was adopted. The ratio between radial force and ring force was established so that the structure would take on maximum curvature with sufficient slope in the radial direction to avoid ponding under snow or rain.

With all of this input established, the final shape under prestress was defined by computer, using an iteration method. This means that, because a set of exact equations cannot be solved for the entire system, simple linear equations were developed for any one point of the net, assuming the rest of the net fixed. The computer then corrected the geometry for one point at a time, moving on to the next one, and repeating this process until the correct geometry of the net was found.

The computer output gave the geometry of the entire structure, all cable forces, cable

For the square-boundaried radial tents designed for the Great Adventure organization in New Jersey, engineer Horst Berger determined approximate curvature and stresses using simple graphic methods to establish equilibrium of forces between rings and radial cables. The drawing below shows assumed forces and calculated shape for one of the radial cables. During erection, the radial cables were jacked upwards, pulling them outwards. This outward action is restrained by the fabric membrane, creating tension in the ring direction. Berger assumes prestress forces based upon what experience tells him would produce reasonable stresses in cables and fabric. The graphic method does not give finite answers, only approximate ones. For actual design, Berger turned to computer solution. Using a cable net as a mathematical model, he can determine the exact shape of tensile roofs, cable lengths, and stresses. From computer solution, space coordinates are determined, which in turn are used with the computer scope to produce pictures.

R radial cable
M membrane
P pole support
C catenary cable
E edge beam

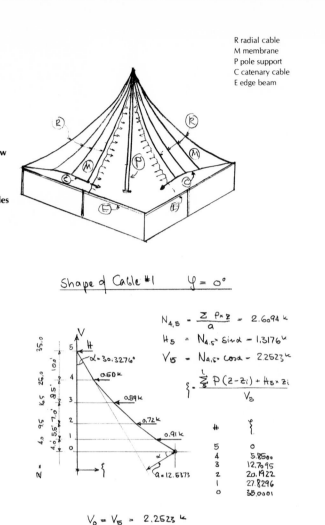

Shape of Cable #1 $\varphi = 0°$

$N_{4,5} = \dfrac{\Sigma P \times z}{a} = 2.6094^k$

$H_5 = N_{4,5} \times \sin\alpha = 1.3176^k$

$V_5 = N_{4,5} \times \cos\alpha = 2.2523^k$

$\zeta = \dfrac{\sum \frac{1}{8} P(2-z_i) + H_5 \times z_i}{V_5}$

#	ζ
5	0
4	5.8500
3	12.7095
2	20.1922
1	27.8296
0	35.0001

$\alpha = 30.3276°$

0.50^k
0.59^k
0.72^k
0.91^k

$a = 12.5373$

$V_0 = V_5 = 2.2523^k$

$H_0 = \Sigma H + P = 4.1376^k$

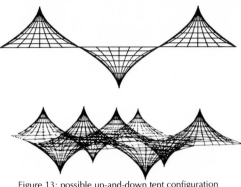

Figure 13: possible up-and-down tent configuration

lengths, and fabric patterns. The geometry and prestress forces were then used as input for the stress analysis under superimposed loads, such as wind and snow. As part of this analysis, the prestress forces can be adjusted so that slackness (loss of tension) is avoided under superimposed loads. Because the shape determined by this process is independent of the prestress level, an adjustment of the prestress level does not require a re-run of the geometry program.

In this case, because of the use of the fabric as the ring system, an additional process of computation was needed to determine the actual fabric stresses in the two-way fabric membrane, which is somewhat more complex than a linear cable system.

Some interesting methods of deriving more complex shapes involve media such as

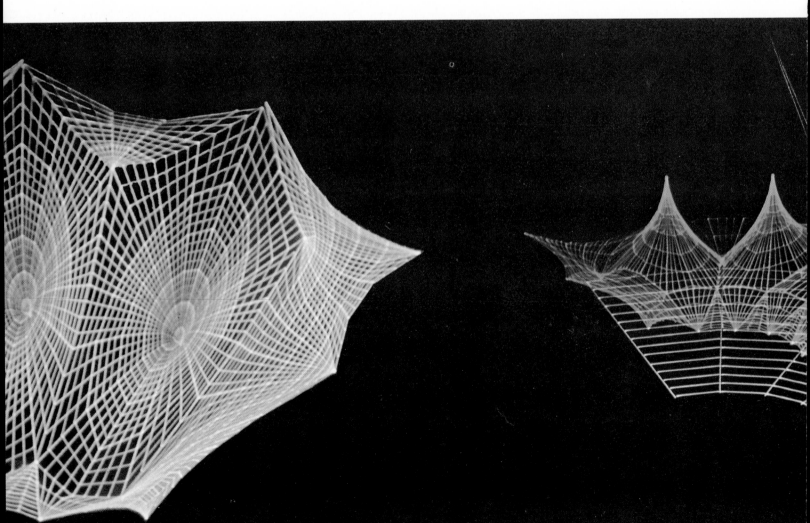

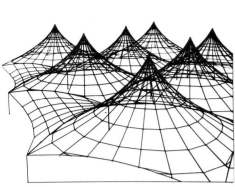 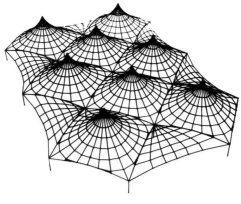

Figure 14 and Figure 15: tent configuration suggested for warehouse application

Figure 16: four-tent configuration for tennis courts

soap bubbles, stretched fabrics and other modeling materials. But these methods are time-consuming, their results are approximate, and soap bubbles, at least, bear little resemblance to common construction. A computerized mathematical procedure, like that described, is therefore an indispensable tool for the design of these structures.

In the tents built at Great Adventure (Figure 12), the radial cables are ⅝-in. bridge strand; the function of the ring cables is taken by the fabric, a vinyl-coated polyester with an ultimate strength of 400 lbs per in. The edge beam is a 14WF steel beam. These four tents, erected in the spring of 1974, are permanent buildings, and are designed to withstand full snow and wind loads.

The program used for the Great Adventure tents developed the shape and provided the cable lengths and the patterning dimensions for the fabric. Stressing was performed with jacks at the top of the mast. At the correct pre-stressing force, the tops of the tents came within ⅛ in. of the design elevation, proving out the accuracy of the design.

Developed system can be applied to multiple tents of complex form

This design system opens the way to many new shapes. Figure 13 shows a computer scope picture of a structure arrived at by stringing together square tents arranged alternately with high and low support points. The addition of edge catenaries—their shape also found by the same computer operation—permits groupings of multiple tent structures such as the proposed warehouse building illustrated in the computer-scope perspectives of Figures 14 and 15.

The arrangement of four square tents into a roof structure for tennis courts (Figure 16) demonstrates another application of the same principles and design tools. The supports are arranged to achieve as close a balance of prestress conditions as possible.

Several other applications are in various stages of development. The most advanced among them is the roof structure for architect William Morgan's Amphitheater at Interama in Miami (Figures 17, 18 and computer scope drawings at the bottom of these pages). Its function is to protect 6,000 theater seats against sun and rain, while retaining an open view and a natural flow of air. This structure had to be designed to withstand 130 mph hurricane winds and, because this is a commercial enterprise, to stay within a tight budget.

The high supports are created by five

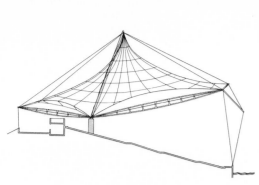

Figure 17: frame and cable supports for Interama

Creative Photographic Service, Inc.

Figure 18: model of Interama Amphitheater

A-frames arranged along an arc. The low supports are identical with the support points of the A-frames. The outer edges have elevations between high and low supports and are held in position by vertical tiedowns, by masts originating from the low supports, and by cables leading to the top of the A-frames (Figure 17). All lateral forces therefore balance out at the low supports, which rest on piers, and the total framework of A-frames, masts and main cables forms a stable support system for the radial tent structure.

The tent structure has a typical spider-web-like configuration (below). The main module, occurring five times, has a pentagonal shape, and the rear module is close to a triangle. The shape of the side modules—the "wings" at each end—is slightly more complex.

The extreme wind loads along the coast of Florida produce stresses that prohibit the use of fabric for ring forces, and thus require cables in both directions of the net. The computer scope pictures show the actual cable configuration. The fabric is a Teflon-coated fiberglass with a strength of 600 lbs per in., and 8 per cent translucency.

The development of structural designs such as these has brought lightweight tensile structures within a cost range that is highly competitive with other low-cost buildings. The availability of excellent fabrics, some of them highly durable and fire-resistant, makes permanent structures possible. The high translucency of these fabrics eliminates the need for artificial lighting in the daytime, and the tent shape is excellent for natural ventilation.

And the complex, flowing surfaces are inherently beautiful.

Figure 3—U.S. PAVILION, OSAKA, JAPAN; Architects: *Lewis Davis, Samuel M. Brody and Alan Schwartzman;* Engineers: *Geiger Berger Associates, P.C.* (structure). *Figure 4*—WILTON TENNIS COURTS; Architect: *A. Robert Faesy, Jr.;* roof structure: *Geiger Berger Associates, P.C. Figure 5*—FLEXIBLE ARCH DOME; Design and engineering: *Geiger Berger Associates, P.C.:* constructed with assistance of students at the Columbia University School of Architecture; patent pending. *Figure 6*—PRESTRESSED FABRIC SHELTER; Design and engineering: *Geiger Berger Associates, P.C.;* patented. *Figure 12*—GREAT ADVENTURE MERCHANDISE BUILDING, NEW JERSEY; Design and engineering: *Geiger Berger Associates, P.C.:* Associate architects; *Hankin and Hyres,* Architects. *Figure 16*—FOUR-POINT TENNIS TENT; Design and engineering: *Geiger Berger Associates, P.C. Figures 17, 18*—AMPHITHEATER FOR INTERAMA, MIAMI; Architect: *William Morgan Architects;* Special consultant, roof design and engineering: *Geiger Berger Associates, P.C.*

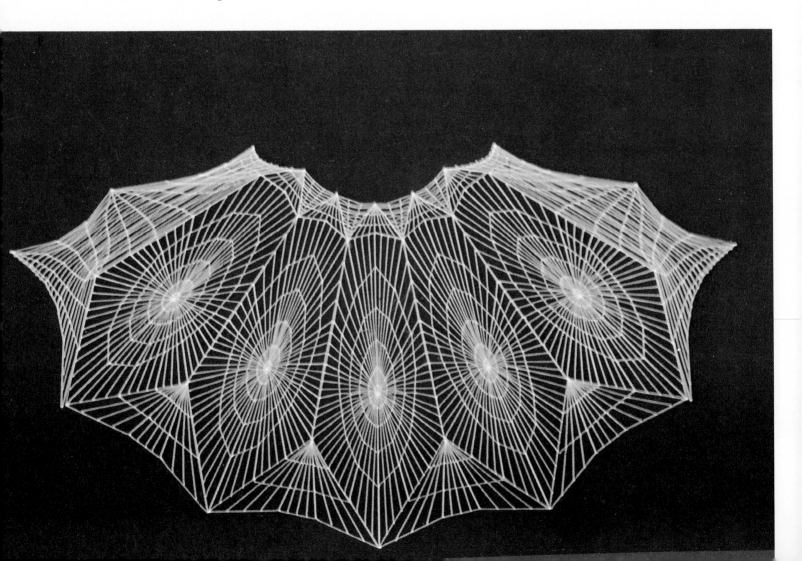

Fabric roof for a California store

Paul Heidrich, chairman of Bullock's of northern California, told Virgil Carter of EPR architects and planners that he wanted ''something more interesting and economical than a black box requiring tremendous amounts of energy'' for the company's new store in San Jose. Presented this challenge, EPR developed their design based upon the following guidelines: 1) the store should be based upon interior merchandising concepts; 2) the store should find new ways to conserve energy; 3) the store should act as a prototype for ''breakthrough'' solutions, developing a completely new type of shopping environment suitable for much larger scale.

Early in design development, engineer Horst Berger suggested several different means of support for a fabric roof, including a four-peaked tent solution and cross arches. Initial cost studies indicated that arches would be less expensive, and, at the time, the owner had reservations about the visual dominance of a pole-supported solution. But now, after enthusiastic customer response at the San Jose store, Bullock's is going ahead with an eight-peaked tent for another large store in the Bay Area.

An advantage of the cross-arch solution is that the arches are stable themselves, and the fabric can ride free over their top surfaces. No cables were needed except edge catenaries which help stress the fabric evenly when jacks are applied.

Most of the roof has two layers of fabric—a somewhat conservative response to thermal, acoustical, and fire-protection concerns—whose varied geometrical pat-

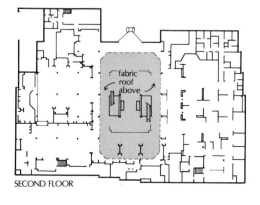

SECOND FLOOR

terns and light and darker translucencies provide counterpoints of visual interest. The double layer gives added thermal insulation as a measure to avoid condensation, though the engineers think this can be avoided with proper ventilation. The inner layer is a different fiberglass fabric than the outer layer, and is sound absorptive. The inner layer also gives a second line of defense against burning embers carried by wind (fire officials required the fabric be subjected to the burning brand test used in California, though the engineers feel the test is not meaningful as the brands slide off the surface at angles of 16 degrees or more, according to Berger). Fire officials also thought that the double-fabric layers would trap more heat in the vicinity of sprinklers so they would respond faster in case of a fire from the inside.

On the energy side, the reflectivity of the fabric helps reject solar heat from the outside, but its translucency (16 per cent for a single layer and 7 per cent for a double layer) allows high lighting levels and a vibrant interior appearance. In the climate of northern California, stores could function with little energy for heating and cooling, according to studies made by the engineers.

Cost of fabric roofs, in place, today averages about $16 per sq ft, in contrast to $9 per sq ft for a conventional flat rectilinear steel frame. But, say the designers, in addition to the energy and marketing values of the fabric roofs, advantages include being able to provide mezzanine space, having less exterior wall surface, and having no built-up roof.

BULLOCK's OAKRIDGE, San Jose, California. Owner: *Bullock's of Northern California.* Architects: *Environmental Planning & Research, Inc., Virgil R. Carter, project architect.* Fabric roof designer, structural engineer and mechanical consultant: *Geiger Berger Associates, P.C., Horst Berger, principal-in-charge.* General contractor: *Swinerton & Walberg Co.* Fabric roof fabricator: *Birdair Structures, Inc.* Fabric roof contractor: *Owens-Corning Fiberglas Corporation Fabric Structures Unit.*

Joshua Freiwald

The undulating fabric membrane structure hovering over Bullock's Oakridge department store, and covering one-third of the roof, gives the store unusual distinction in a San Jose, California, shopping center. The application of the membrane roof to this 150,000-sq-ft, two-level store is said to be the first for a retail structure in this country.

Bill Apton

Structural support for the fiberglass fabric membrane is two pairs of crossed, laminated arches that rise 22 ft, span 96 ft, and are 32 ft apart. Hand-operated winches hoisted the single-piece, 18,000-sq-ft fabric membrane up over the arch frames. To prevent abrasion of the fabric, a strip of the same material was placed on the top surface of the arches. Originally, five pairs of arches were considered, but to cut costs, the structural engineers, Geiger Berger, suggested eliminating three of five cross-arches and increasing the capacities of the remaining two. There are no cables in the fabric except edge catenaries.

The membrane was stressed by means of hydraulic jacks. It was attached to an encircling concrete curb with a special two-plate clamping system.

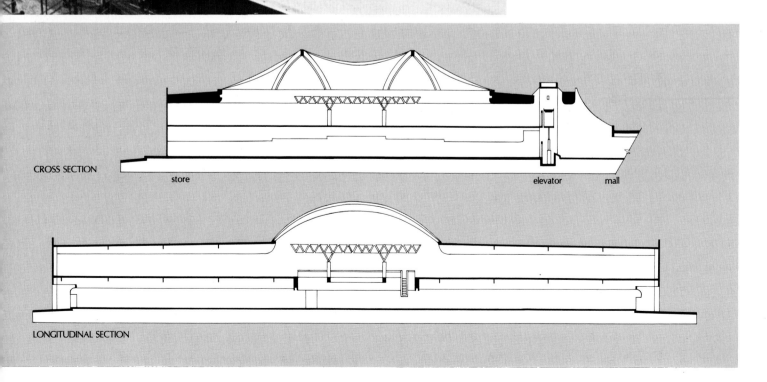

CROSS SECTION

store   elevator   mall

LONGITUDINAL SECTION

Pavilion tents for the Bicentennial

To house a variety of celebrations throughout Philadelphia's Bicentennial summer, architects H2L2 conceived a series of emblematic and festive "tents" placed around the city. Three of the series are shown here: the Folklife Pavilion in Eakins Oval (at top and lower left), one of a group of food-service pavilions (lower right), and the Independence Mall Pavilion (overleaf).

The undertaking of so many and such varied tension structures afforded engineer Horst Berger an unusual opportunity to demonstrate the theories and designs of tensile fabric structures which he has developed in recent years. Since describing the engineering discipline that must underlie the design of tent structures (see pages 29-36), Mr. Berger has considerably refined and extended mathematical procedures for determining cable and fabric stresses and for deriving structural forms. These sophisticated procedures can achieve continuous curves in the radial plan, allowing a wider range of possible shapes than earlier programs.

The "force-density" method, adapted by Geiger Berger from a German program for cable nets, is used to determine the shape of prestressed membrane structures: working with fabric study models—and their own judgment and experience—the engineers provide as input for the program approximate length and approximate force for each member, and receive as output the exact location of each node (that is, the intersections of radial and concentric forces) and the exact length and force for each member. The structural analogy offered by the engineers is that of a net of rubber bands that approximately fits over a number of support points. As the net is stretched to connect the support points, each rubber band changes its length and stress to fit the geometry required, and the net thus defines a shape in which every node is in equilibrium. When the exact shape of the membrane is determined, the engineers can specify the length and prestressing of each cable and the fabric patterning. The program may be run several times to arrive at the proper structural shape and uniform stress distribution.

BICENTENNIAL STRUCTURES, Philadelphia. Owner: *Philadelphia '76*, a city agency. Architects: *H2L2, Architects/Planners—Paul C. Harbeson (partner-in-charge), Barry N. Eiswerth (director of design)*. Engineers/design consultants: *Geiger Berger Associates, P.C.* (structural). Fabricators for structural fabrics: *Birdair Structures* (Independence Mall and Folklife Pavilions); *Air-Tech Industries* (food pavilions). Contractor and erector: *Owens/Corning Fiberglas Construction Division* (Independence Mall and Folklife Pavilions).

The Folklife Pavilion at Eakins Oval consists of two rows of what are in effect half tents, together spanning 68 ft. Supported by 55-ft vertical masts, the structure comprises two rows of radial tents, with fabric cut away outside the masts. The fabric membrane is constructed essentially of parallel flat strips, stressed by cables only at the ridges (mast to mast), valleys (ground to ground) and the edges; radial cables were eliminated. Repetition of panel shapes substantially reduced fabrication costs.

The GB Shell—a tensioned fabric structure patented by Geiger Berger—was used for a number of food-service pavilions. The modular structure consists of eight identical flat fabric elements, which are basically triangular with curved edges. The frame is constructed of 12 pipes of equal length, which equalize prestress forces; foundations are thus required only to resist such superimposed loads as wind and snow. Here, modules were mounted on 4-ft concrete posts to place fabric above the reach of vandals.

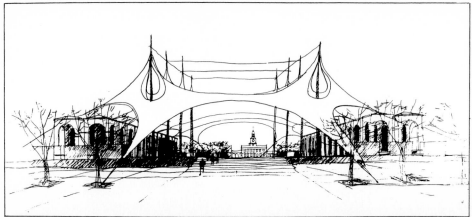

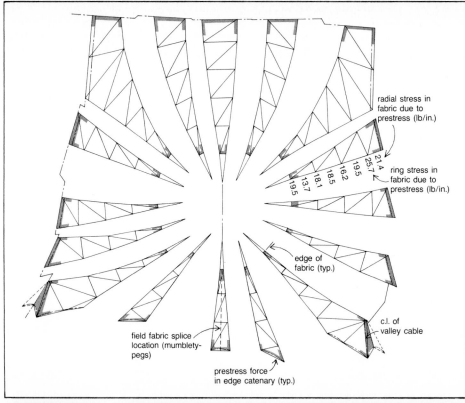

radial stress in fabric due to prestress (lb/in.)

ring stress in fabric due to prestress (lb/in.)

21.4
25.7
19.5
16.2
18.5
13.7
19.5

edge of fabric (typ.)

c.l. of valley cable

field fabric splice location (mumblety-pegs)

prestress force in edge catenary (typ.)

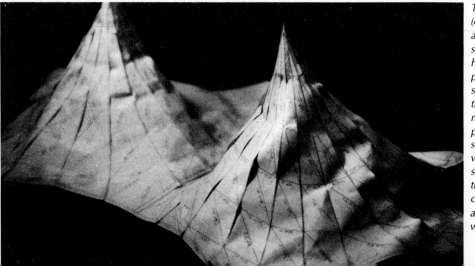

The steps in the design process were well documented (column left) for the Independence Mall Pavilion. The architects first suggested a tent structure that would span both sides of an existing arcade, and that would have sufficient height to allow a clear view of Independence Hall. The engineers then constructed a study model (masts were later tilted to ease forces on the valley cable anchorages when the entire tent was moved inside the arcade). The fabricators received a pattern that specified both dimensions and fabric strengths. The paper "model," lower left, assembled with a cut out pattern drawing, is a pragmatic demonstration of the actual shape of a tent (not, in this case, the Independence Mall Pavilion). Because cables come to the peak at different angles, the connector is asymmetrical to bring them to the same working point with equal forces.

Efficient shapes for air structures

Since designing the air-supported roof for the U.S. Pavilion at Osaka's Expo '70, engineers Geiger Berger Associates have designed a number of more advanced pneumatic structures and have greatly developed the theoretical bases for these buildings. (Mr. Geiger has, indeed, patented a mathematical system for determining the relationship of cable location to ring plan.)

One of the basic rules for the design of large-scale air structures is that they assume as flat a surface as possible to minimize the effects of wind load. But the restraint on the cables needed to produce this low profile subjects the compression ring to large forces. At Osaka, the ring is a superellipse—a sort of squashed oval whose form is mathematically defined. Chosen for functional and esthetic reasons, this shape also turned out to be most efficient structurally. Superellipses are especially effective when cables are strung from the "corners" so that the ring acts as an arch, thus reducing bending stresses.

In recent designs, Geiger Berger have been able to reduce the number of cables—and hence the number of expensive connectors—required for air-supported roofs. An essential limitation in spacing these cables is the strength of the fabric, which can span no more than 40-45 ft. The great number of cables at Osaka (see comparative plans below) reflects the extraordinary wind-load strength needed in a typhoon zone. At Santa Clara, however, where neither high winds nor snow were major considerations, the number of cables is reduced to three in each direction.

Leavey Center at the University of Santa Clara, shown on these pages, comprises two air-supported structures—one a superellipse for basketball and other sports, the other a smaller superellipse for swimming meets. The arena roof, which is permanent, is Teflon-coated fiberglass restrained by skewed cables. The swimming pool roof is retractable. The fabric—vinyl-coated polyester—is stored in a sausage-like roll and will be stretched and connected to the ring manually; transverse cables are encased in the fabric.

THOMAS A. LEAVEY ACTIVITIES CENTER, University of Santa Clara, California. Architects: *Albert A. Hoover and Associates* (principal); *Caudill Rowlett Scott* (design); *Philip Welch* (consultant). Engineers: *Pregnoff/Matheu/Kellam/Beebe* (structural); *Geiger Berger Associates, P.C.* (air structure); *G.M. Simonson & T.R. Simonson* (mechanical/electrical). General contractor: *Johnson and Mape Construction Co.* (construction manager).

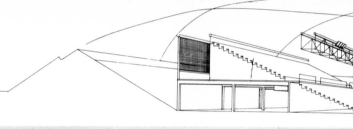

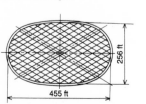

U.S. Pavilion, Osaka, Japan

455 ft · 256 ft

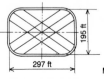

University of Santa Clara, California

297 ft · 195 ft

Milligan College, Tennessee

212-ft diameter

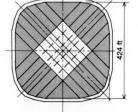

Uni-Dome, University of Northern Iowa, Cedar Falls

424 ft

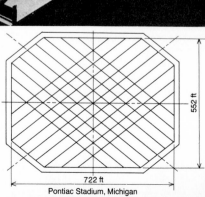

Pontiac Stadium, Michigan

722 ft · 552 ft

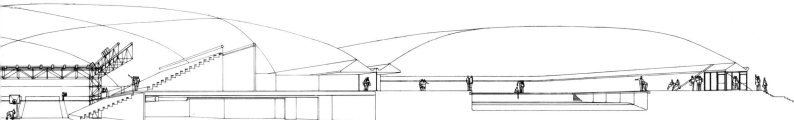

At Santa Clara sports facility, conventional pipe hangers support 1⅞-in. cables below fabric, which is joined by bolts through continuous neoprene strips. Heat-sealed fabric seams, 12 ft apart, are perpendicular to long edge of ring. Where cables cross, they are held by U-clamps welded to steel-pipe sleeves.

The main building in the new athletic complex will seat 5,000. Because the fabric roof cannot support heavy mechanical and electrical equipment, freestanding frame accommodates lights and other gear; power lines are carried above roof cables. (Wrinkles apparent above disappeared when inflation was complete.)

William C. Eymann

Air roofs for two large sports stadiums

In September and November 1975, the two largest cable-restrained, low-profile air-supported roofs were inflated: Pontiac (Michigan) Metropolitan Stadium, covering 10 acres of clear-space, and UNI-Dome at the University of Northern Iowa, covering 4.1 acres. For scale comparison, the U.S. Pavilion at Expo '70 in Osaka was one-fourth the area of Pontiac.

Substantial savings in roof construction costs and time were achieved with Pontiac and UNI-Dome. Cost of the roofs, including compression rings, was $11-$12 (1974) per square foot of enclosed area. The roof for UNI-Dome took only four weeks to erect. The Pontiac Metropolitan Stadium roof, more than twice the area, took close to four months, though it could have gone faster if there had not been interruptions resulting from the Detroit Lions' games.

Sophisticated engineering and sophisticated materials made these domes possible

These permanent structures could not have been built at the costs they were without the engineering principles developed by engineer David Geiger or the fabric and panel construction developed by a collaborative group of manufacturers. A unique structural concept was first demonstrated in the Osaka structure based upon the principles of skewed symmetry—whose application to air-supported and arched structures has been patented by David Geiger. With a skewed criss-crossing of cables rather than a rectilinear grid, and the use of superellipse geometry for the compression ring (obtained with exponents larger than 2.0 in the ellipse equation), cable tonnage can be reduced by one-third, and bending of the compression ring caused by pressurization force on cables can be reduced to zero. Some bending can occur when there are unbalanced cable forces caused by wind loads, but application of skewed geometry substantially reduces ring costs.

After Osaka a new fabric was developed for greater permanency. The Osaka structure was covered with vinyl-coated fiber glass. Because ultraviolet rays degrade vinyl, the life expectancy of the fabric is about 10 years. The new fabric—*Fiberglas* fabric coated with *Teflon* fluorocarbon resin—has a number of properties essential for permanent structures:

UNI-Dome, University of Northern Iowa

Robert E. Fischer

Pontiac, Michigan Metropolitan Stadium

Birdair Structures, Inc. photo

long life (estimated to be more than 25 years), fire-safe attributes, and high strength. The fabric will not support combustion (tested by the ASTM Oxygen Index method, an atmosphere of more than 95 per cent oxygen would be required to sustain combustion).

One unusual property of the fabric is that dirt does not stick to it, so it is self-cleaning. Another unusual property is its translucency, which can be varied from opaque to 14 per cent light transmission (reflectance is 75 per cent and absorption, 11 per cent), Much of the time, therefore, electric light is not needed.

Two different formulations of *Teflon* are used, the final coating allowing high-strength, heat-sealed seams for fabricating larger panels from 12-ft wide material. Strength of the 0.032-in.-thick fabric permits spans of 40 ft or so, but panels may be as long as is convenient for construction (at Pontiac, the longer panels ranged from 115 to 190 ft). The fabric for Pontiac and UNI-Dome was woven, coated and pre-stretched by Chemical Fabrics Corporation, and was made into panels by Birdair Structures, Inc.

Apparently, stability must be considered during construction as well as afterward

Many safeguards have been built into the design of these air structures to make sure they will withstand the rigors of weather and be fail-safe in terms of human occupancy. What was not foreseen before these two structures were built, however, were problems—some merely a nuisance, others potentially more serious—that could occur with construction not fully completed. Experience with harsh November weather in Cedar Falls, Iowa, and Pontiac underscored the fact that the pneumatic system (fans and air-tight structure), the snow melting system, and backup system (standby power) need to be working as designed.

At the uncompleted UNI-Dome, for example, some fabric panels had to be replaced because during a violent storm on November 9, which included tornado activity, lightning ripped one panel, and the roof deflated during a power outage (standby power was not yet in operation and lightning arresters had not yet been installed). Before the panel could be replaced, 50-mph winds later in the week damaged seven more rectangular panels as they were buffeted. On the other hand, the *inflated* roof at Pontiac stadium took gusts as high as 75 mph from the same wide-area storm with no trouble at all.

Later that month, though, on Thanksgiving Day, a wet snow at Pontiac caused several panels to invert, and melted snow dripped through the drain holes in the roof onto the playing surface. The primary reason, according to the engineers, was that the mechanical system was incomplete and the 10 direct-fired blowers used for snow melting were inoperative. The four indirect blowers (which only re-circulate air) worked but could not, alone, prevent snow buildup. In addition, the internal pressure could not be increased to design level because some garage-type doors in the building need to be reinforced for the 12 psf top pressure intended for snow loads.

There are four modes of operation for cop-ing with snow loads: 1) the snow can be melted, 2) the pressure can be increased up to the maximum of 12 psf, 3) if these two modes fail, individual panels can deflate and emergency drains located only over the playing field open, 4) if still greater load accumulates due to snow, the roof can handle loads in a deflated condition (12 psf for Pontiac and 30 psf for Iowa).

Experience in erecting the roofs of these two stadiums has shown that during construction high winds buffeting the long panels can damage them (diamond and triangular panels have not been affected.)

Because pressurization gives stability, Geiger intends to require in new projects that air pressurization and standby power systems be fully operable before any panels are erected. Furthermore, to ensure roof stability both during erection and in the remote event of freak, catastrophic damage, the long-rectangular-panel areas will need to be pressurized during panel erection and in case of openings that could result in deflation. This will be accomplished by "buttoning up" the acoustical panels on all edges so that the two skins can be inflated by fans.

Lower first cost made the difference in having a roof over one's head

Architects for Pontiac Metropolitan Stadium, O'Dell/Hewlett & Luckenbach, Inc., turned to the air-supported roof as the only viable alternative for a weather cover in view of the limited budget. The original price for a stadium with a steel-vault roof was higher than allowable financing. Furthermore, subsequent legal entanglements on financing delayed the bond issue while building costs kept inflating.

Having given up on a conventional roof, the architects initially proposed an air-supported roof as a future add-on roof. But soon after, the city of Pontiac lent the stadium authority additional money so the roof could be included in initial construction.

The basic stadium design was modified accordingly. A compression ring was incorporated into the design, and the plan shape was adjusted from 45- to 37-degree corners so that the principles of skewed symmetry could be applied. Even so, some bending results in the ring because the plan has truncated corners and is not a superellipse (though the so-called funicular curve, or pressure line, does fall within the width of the compression ring). The design for bending of the compression ring made it cost somewhat more per square foot of plan area than UNI-Dome. A compensating factor, however, was the ability to mount the blowers with their direct-fired heaters on the ring platform.

Initial cost also was an important factor at the University of Northern Iowa because the UNI-Dome is being financed entirely by student fees and by donations. Architects were Thorson-Brom-Broshar-Snyder. Besides serving as a facility for varsity and intramural sports, the UNI-Dome will host variety shows, concerts, conventions and many other events.

Cost of the UNI-Dome roof plus ring is 31 per cent of the building total ($6.5 million in 1974). The ring cost was about $½ million and the roof cost (fabric and cables) about $1.5 million.

Cost of the Pontiac stadium roof plus ring is 10 per cent of the building total ($41.9 million in 1974). The ring cost $1.5 million and the roof cost was $2¾ million. Cost of the ring per square foot of covered area was about 40 per cent higher than the ring at UNI. Because it is 100 ft above grade, it had to be designed so that the formwork would be self-supporting. Plate girders 6 ft feep at the outer and inner circumferences of the ring serve as forms, and also as tension steel for the horizontal bending of the ring.

Cost of the roof for the UNI-Dome is about 25 per cent higher per square foot of covered area than Pontiac stadium because it has a greater curvature and because about 60 per cent of the roof has a double layer. The second layer, also *Teflon*-coated *Fiberglas*, but lighter in weight and porous, is multi-functional. First, it is an acoustical fabric—i.e., it is a sound absorber, having a noise-reduction coefficient of 0.65 as it is mounted. Secondly, it forms a plenum for warm air to melt snow if it collects on the roof. Thirdly, it forms an insulating boundary of air ($U=0.60$).

Just a few horsepower of fan capacity are needed to keep the roofs airborne

When the Pontiac Metropolitan Stadium is unoccupied, only two of the 29 fans are required for air-pressure support. Because the Pontiac stadium is ventilated, not cooled by refrigeration, and because smoking is allowed, a large fan capacity was needed to move sufficient air. For this reason also, the translucency of the panels was kept at 8 per cent, rather than the 14 per cent used at UNI-Dome.

The 80,000-seat Pontiac stadium has twice the plan area and about three times the volume of UNI-Dome. The Iowa stadium with a maximum seating of 25,500 for concerts and convocations has two 40,000 cfm (15 hp) fans that circulate air and maintain pressure when only the field level is in use. Additionally, there are two 135,000 cfm (125 hp) fans that are used during inflation of the roof, for mass exiting from the building, during heavy snow fall for aid in snow melting, and during spectator events for heating and cooling the entire space.

UNI-DOME, Cedar Falls, Iowa. Owner: *University of Northern Iowa.* Architects: *Thorson-Brom-Broshar-Snyder.* Engineers: *Geiger Berger Associates P.C. (structural and mechanical); Flack & Kurtz (electrical).* Consultants: *Ranger Farrell & Associates (acoustical).* Contractors: *John G. Miller Construction Company (general); Owens-Corning Fiberglas Corporation, Construction Services Division (roof).*

PONTIAC METROPOLITAN STADIUM, Pontiac, Michigan. Owner: *Pontiac Stadium Building Authority.* Architects: *O'Dell/Hewlett & Luckenbach, Inc.; Kivett and Myers (consulting architects).* Engineers: *Geiger Berger Associates, P.C. (structural for roof, mechanical for covered stadium); McClurg & Associates, Inc. (structural for stadium); Hoyem Associates, Inc. (mechanical and electrical).* Consultants: *Sasaki, Walker Associates (site planning and landscape architecture); Coffeen, Anderson & Associates, Inc. (acoustical).* Construction manager: *Barton-Malow Construction Management Services.* Roof contractor: *Owens-Corning Fiberglas Corporation, Construction Services Division.*

UNI-Dome, University of Northern Iowa

Area (excl. ring) 168,000 sq ft
Area (incl. ring) 190,000 sq ft
Roof rise 48 ft
Costs per sq ft of plan area inside ring:

ring	$2.89
roof	9.07
total	$11.96

(July 1974)

The dome consists of 29 fabric panels, 25 rectangular and four triangular, and 12, 2⅞-in. diameter stranded steel cables. The roof at the center is 125 ft above the floor. The roof is double-layered for about 60 per cent of the area, the lower layer being acoustical panels that also provide a plenum space for moving warm air under the roof during a heavy snow fall. When deflated the lowest cable is 18 ft above the floor. Flap-covered holes in the roof are for drainage in the deflated condi-tion. Though low spots are determined by computer, location of pooled water varies according to loading.

The compression ring is made of precast sections, the longest being 53 ft. Epoxy grout was used for horizontal joints and vertical joints were poured concrete. The box section is used as the supply-air plenum for the stadium.

The ring is supported by precast double tees, which also serve as the ex-terior wall, and by interior precast col-umns (see detail below).

The wall was not high enough to permit lighting to be installed around the ring and avoid glare, so the fixtures were mounted on steel-frame triangles and hung from the roof cables, though engineer David Geiger would prefer, as much as possible, to avoid concen-trated loads on the cables.

The panels were erected with a crane to hoist rolls of fabric, and with workers guiding the fabric from plat-forms hung from the roof cables and the roof itself.

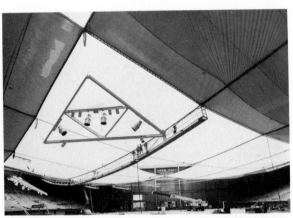

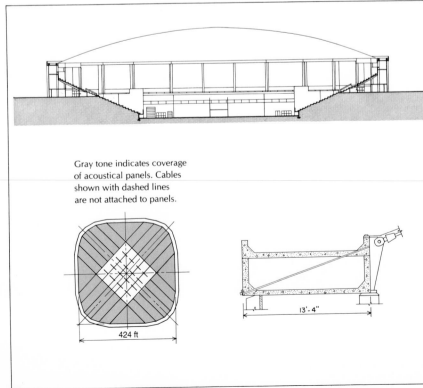

Gray tone indicates coverage of acoustical panels. Cables shown with dashed lines are not attached to panels.

424 ft

13'-4"

Pontiac, Michigan
Metropolitan Stadium

Area (excl. ring) 376,000 sq ft
Area (incl ring) 430,000 sq ft
Roof rise 50 ft
Costs per sq ft of plan area inside ring:
ring	$4.00
roof	7.27
total	$11.27
	(Feb. 1974)

The dome consists of 100 fabric panels—64 diamond-shaped, 32 rectangular, and four triangular—and there are 18, 3-in. diameter steel cables. The roof at the center is 202 ft above the floor. The roof has a single layer; sound absorption is provided by tufted acoustical baffles hung vertically from the cables.

The compression ring consists of 6-ft deep plate girders and poured-in-place concrete, which is prestressed transversely to obtain horizontal shear transfer from the concrete to the plate girder shear studs. The ring is supported by steel columns and angled struts.

The fabric panels were installed in two phases. The diamond-shaped center panels, in rolls, were hoisted by movable crane. Workmen used U-shaped clamps to attach them to the cables. Panels have nylon-rope edges. First the U-clamps were bolted to the cables. Then the panels were clamped between a sandwich of neoprene waterproofing strips and aluminum plate strips at the roped edges.

The diamond panels were hoisted by a movable crane alone. Hoisting of the long panels was accomplished, on the other hand, by use of two pairs of unique, movable A-frame towers. The towers traveled on rails mounted atop the compression ring. In each pair, one tower had a gasoline-engine winch, while the other was used as a dummy for balance; a movable hoist was installed on cables strung between opposite towers.

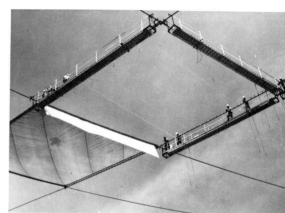

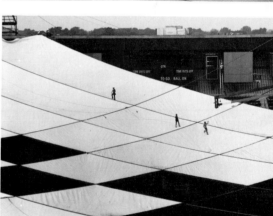

Owens-Corning Fiberglas photos

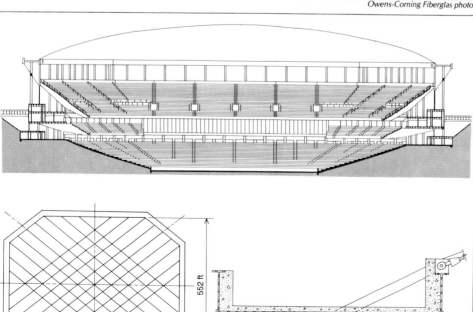

CHAPTER 3:

Structures– Systems and Prefabrication

This chapter gives examples of some rather sophisticated applications of prefabrication in concrete and steel, illustrating how engineers are taking advantage of the economies of factory-assembled components to produce buildings tailored to specific architectural requirements.

In adapting European housing systems to United States room sizes and code requirements, the engineers ran into a number of difficulties. In essence, they found that they had to do more extensive work than they expected—in making adaptations and checking hundreds of shop drawings—to use a European system for a one-off project. It became clear that the economies of this approach lie in being able to produce many carbon copies of the same design. On the other hand, a technique such as "flying forms" for concrete, widely used in Europe, appeared to work well for one-off projects so long as they involved a lot of repetitive modules, and the construction sequence could proceed smoothly.

The use of factory-built boxes for a Texas hotel speeded construction on a restricted-access site, but the approach, except for a few isolated examples, has not caught on—continuity of market and the acceptance of standardized units being prime requisites for success.

In custom design, however, two California engineers demonstrated they could save money and time by having steel space trusses fabricated in convenient-length modules in a shop. Detailed so that they could easily be bolted together in the field, they were utilized in buildings as diverse as a savings and loan bank, a library, and a ski lodge.

Precast multistory space frame

Architect and engineer, combining their special skills, develop a building system that is versatile, easy to erect, and structurally efficient—and in its first time out has saved real money.

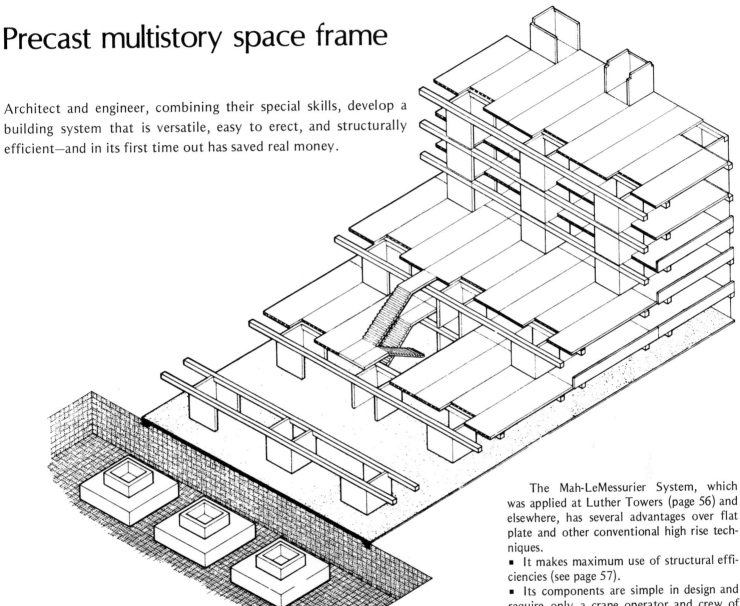

The Mah-LeMessurier System, which was applied at Luther Towers (page 56) and elsewhere, has several advantages over flat plate and other conventional high rise techniques.

■ It makes maximum use of structural efficiencies (see page 57).

■ Its components are simple in design and require only a crane operator and crew of four to erect.

■ Components can be fabricated in a plant or on-site by any competent contractor.

■ Esthetic options are not severely limited by component design.

■ Savings in construction time up to 50 per cent have already been realized. This means earlier occupancy and, in the case of rental buildings, earlier cash flow.

■ Savings in construction cost have ranged between 8-12 per cent for low rise; 18-27 per cent high-rise over flat plate construction because flat plate is more complicated to build. Reinforcement has to be installed in the field and a lot is required for the columns. The precast system, by contrast, uses its structure efficiently and reinforcing is done in the plant. Post-tensioning, in the field, is simple and inexpensive.

When the Memphis Housing Authority, under instructions from HUD, directed the firm of Walk Jones and Francis Mah, Inc. to redesign and take new bids on their apartment project for the elderly, the architects turned frustration into serious research. The firm began to examine prefabricated concrete building components and assembly procedures with a view toward simplification. From this research, which lasted several years, Mah developed the nucleus of a building system that he took to William LeMessurier, Boston structural engineer. LeMessurier quickly saw the system's potential and sought ways to exploit it.

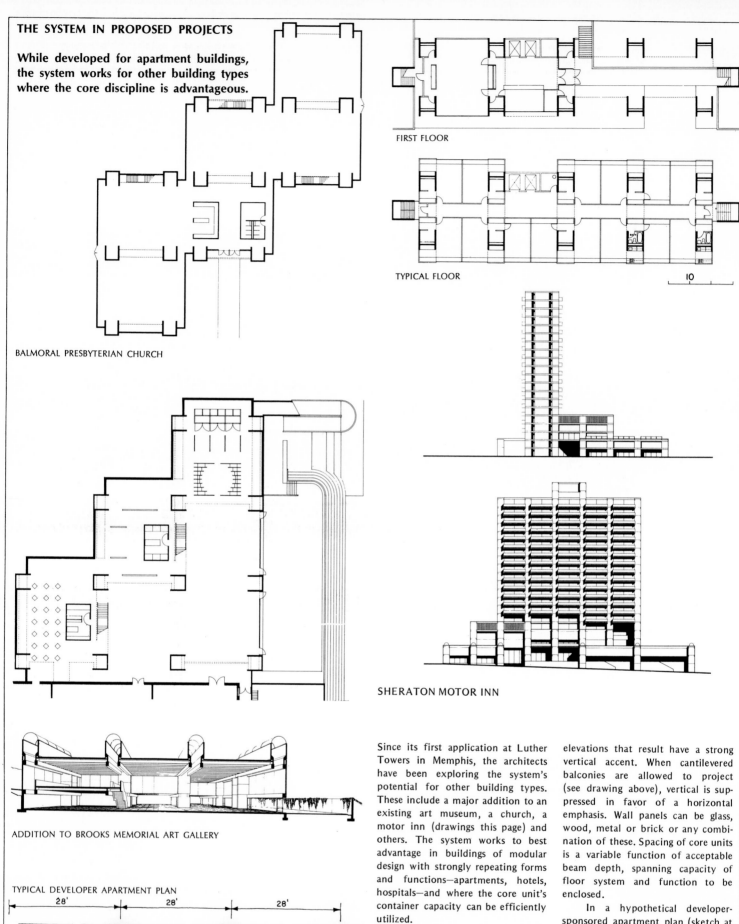

THE SYSTEM IN PROPOSED PROJECTS

While developed for apartment buildings, the system works for other building types where the core discipline is advantageous.

BALMORAL PRESBYTERIAN CHURCH

FIRST FLOOR

TYPICAL FLOOR

10

SHERATON MOTOR INN

ADDITION TO BROOKS MEMORIAL ART GALLERY

TYPICAL DEVELOPER APARTMENT PLAN

28' 28' 28'

L.R. BR. L.R. BR. BR. STUDIO

28' 36' 20'

1 BEDROOM 2 BEDROOM STUDIO

Since its first application at Luther Towers in Memphis, the architects have been exploring the system's potential for other building types. These include a major addition to an existing art museum, a church, a motor inn (drawings this page) and others. The system works to best advantage in buildings of modular design with strongly repeating forms and functions—apartments, hotels, hospitals—and where the core unit's container capacity can be efficiently utilized.

The system's most structurally efficient use is in buildings that range between 5 and 25 stories—the zone in which the concrete works most intensively. With design modifications, this upper limit can be extended. Below 5 stories, the economic advantage over conventional construction is sharply reduced.

When the service cores are stacked on the outside wall, the elevations that result have a strong vertical accent. When cantilevered balconies are allowed to project (see drawing above), vertical is suppressed in favor of a horizontal emphasis. Wall panels can be glass, wood, metal or brick or any combination of these. Spacing of core units is a variable function of acceptable beam depth, spanning capacity of floor system and function to be enclosed.

In a hypothetical developer-sponsored apartment plan (sketch at left), service cores have been moved from the outside wall to a position next to the corridor; the mechanical chases allow back-to-back plumbing. In this example, the structure at the outside wall need only resist gravity forces because the service cores and their floors provide a moment-resisting frame to take the wind loads.

1. System components: cores, beams, double tees and wall panels

The basic components and the details were kept simple, but they perform sophisticated functions.

2. Transportation of components to the site

3. Early construction stage, wall panels hanging from cores

4. Double tees being placed for one of the floors

The "working" structure that takes both gravity and wind loads is comprised of the "U"-shaped service core units, supporting 24-ft-long beams which carry 38 ft. 6 in. double tees. A very effective space frame is obtained from precast elements by making one of them—the core unit—of a much larger scale than conventional vertical supporting elements. The 8-ft-wide core units fit on the flat bed of a truck. Their size, however, was determined by the length of a bathtub plus the width of space required for a mechanical chase. The double tees in Memphis came 8-ft wide, but floor structural elements could just as well have been 4-ft wide, fitting the modular dimension of the core.

The structure goes together as simple as one, two, three. The core units are grooved to engage the top of the T-shaped wall panels, and are notched for the beams to go through. The core units are post-tensioned vertically, clamping beams between them.

5. A service core being eased into place

6. Cores in place; grooves on sides slide over walls

7. Placing epoxy grout between beam and core

8. Troweling grout prior to setting next core unit

Post-tensioned cores work as one of the three elements in a wind-resistant frame. To ensure that the prestressing force would be distributed evenly, it was necessary for grout to be applied to the top edges of the cores. Metal shims are placed at corners of cores for leveling. Neoprene pads also are used there to prevent local stress concentrations at the shims when prestressing is applied. The pads also keep grout from being squeezed out as core units are set.

Stems of all double tees have steel plates set in them where they rest on the beams, and these are welded to companion steel plates on the top face of the beams.

Core units are prestressed by means of steel rods, the post-tensioning operation being performed every three floors with hydraulic jacks. Rods are joined to one another by means of threaded connections.

Where ends of beams abut, a shear connection is made so that the beams work in a wind resistant frame in the longitudinal direction of the building.

9. Steel plates on face of beam for welding to tees

10, 11. Post-tensioning of core units

13. Luther Towers, Memphis, Tennessee

12. Finished floor with conduit set atop tees

LeMessurier's calculations showed that overturning moments of the cores induced by wind would be too large if the floors were only simply supported. To remedy this, he developed an ingenious arrangement in which the cores, the double-tees and the beams structurally complement one another.

For those double tees that span between core units, a shear connection is made between the ends of the stems of the tees and the core units. When the cores try to bend in the wind, the cores tend to rotate the tees. This rotation throws bending moments into the double tees, a lever arm having been formed between the shear connection and the point of support of double-tee on the beam. By this means only a shear connection is needed at the ends of the tees. In effect, rigid frame action has been achieved without moment connections.

In a somewhat similar fashion, the beams and cores resist wind load in the longitudinal direction.

Structural behavior is shown in the drawings at right.

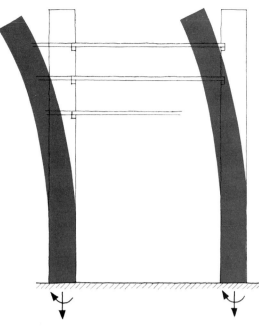

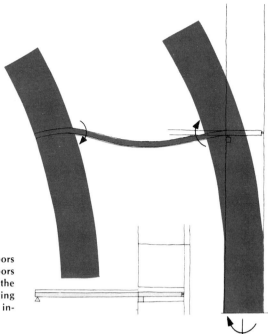

HOW THE STRUCTURE WORKS

The inherent structural capabilities of an efficiently shaped core, beams and tees are exploited utilizing connections that are simple to make and are inexpensive.

Overturning moment of cores is too large when floors are only "simply" supported. Translation of floors takes place (above). But with this new system, the engineer designed restraint into the system by utilizing the moment resistance of the floors (right). Only inexpensive shear connections were required.

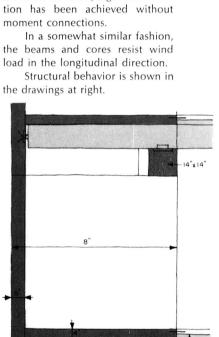

Section and plan of core, beams and ends of tees showing location of prestressing rods, shear connection of tees to core, and bearing of tees on beams.

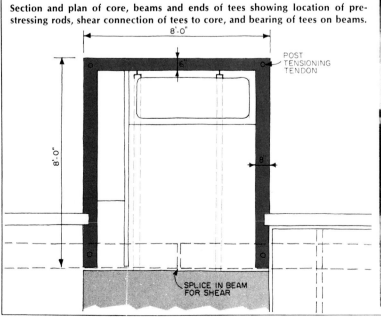

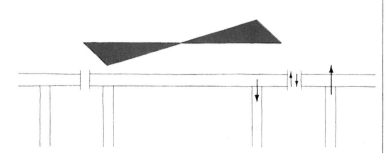

Moment resistance of beams is utilized in the longitudinal direction of the building to provide wind resistance.

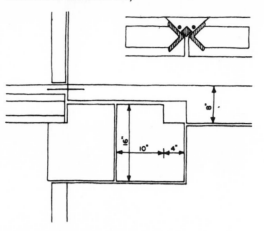

THE SYSTEM ALSO WORKS WELL WITH CONVENTIONAL PLANS

The cores can be moved to the interior for planning reasons, but structurally they function in the same way

Plan of the core unit showing location of holes for post-tensioning rods, and beam shear connections.

Section through center of core unit and beams.

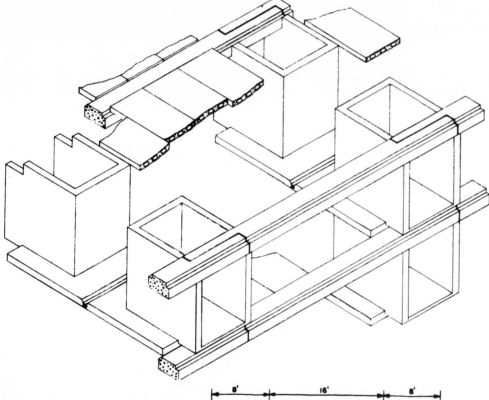

The drawings on the page show some modifications from details presented earlier, intended for application in a conventional 60-ft wide high-rise apartment building, with bathrooms pulled in next to the corridor. The beams now span 20 ft rather than 16 ft, so they are joined in a shiplap configuration to provide greater restraint for gravity loading. Because the span of floors between beams is less than 20 ft prestressed hollow-core slabs or precast slabs with ordinary reinforcement can be used. The same is true of span between core beams and spandrel beam at the outside wall.

In this modified design, the post-tensioning rods are spaced equidistant from the centroid of the core, avoiding eccentric loading of the core.

In this suggested design the cores have a larger area floor slab poured integral with them, forming the floor slab of the corridor.

Resistance to wind load is provided by the cores in conjunction with the bending resistance of the floor slabs of the core. Slab connection detail is shown inset in right-hand drawing above.

The cost picture

Cost studies have shown that the structure for this system can run from 15 to 25 per cent less than that for flat plate construction—the most commonly used concrete system for high-rise apartments.

Structural elements of the system for a conventional high-rise apartment building. Note that core is notched to receive the ends of contiguous beams. Floor slab poured integral with the core extends out on either side of the core and also extends back.

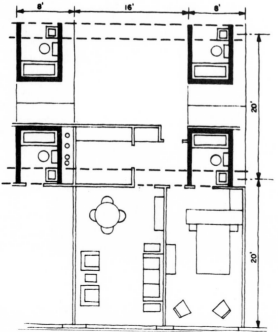

Adaptation of a European housing system

European industrialized building systems for housing cannot be used directly in this country without some modifications—for some a few; for others a great many. The result is that architects and consulting engineers have more to do the first time they work with a system than they would with traditional construction. At the least, they have to familiarize themselves with many details of the particular industrialized system they will be working with.

Modifications to the transplanted industrialized systems may be necessary to accommodate differences in code requirements between Europe and the U. S.; differences in structural design approaches and field practices; differences in materials and practices acceptable to sponsoring and lending agencies (such as HUD and FHA) and to regulatory organizations (such as Underwriters' Laboratories); and, finally—hardly an insignificant factor—differences in what people expect in the way of standard of living—in room sizes; mechanical, electrical and plumbing systems, etc.

The architect may need to analyze the building with respect to how many different types of concrete panels are necessary—different sizes and shapes of panels with different sizes and shapes of openings. Of course, the fewer types of panels there are, the better will be the cost picture. The structural engineer will need to understand that there are differences in such things as reinforcing details between conventional construction and industrialized buildings (because most of the structure is factory fabricated, reinforcement is placed more accurately than in the field; concrete strengths are more uniform, etc.) The extent of the structural engineer's participation will depend upon whether he has been engaged to adapt a European system to American practice, or whether he is operating in the traditional fashion of consultant to an architect in using an industrialized system that has already been adapted by the entrepreneur.

In the first case he could get involved in redesigning the reinforcement (an example shown in this article) for more efficient use of the material.

In the latter case he will, no doubt,

This 20-story apartment building in Yonkers, New York was one of the first applications of a European industrialized housing system in this country. Panels were trucked a short distance to the site from a factory in the Bronx. Maximum dimension of floor slabs is 12 ft (spanning distance) by 25 ft, and walls are a maximum of 32 ft in length. These limitations were set to fit the size of casting beds and to meet transportation requirements. Maximum panel weight was limited to 10 tons for handling purposes. Average rate of placement was 1½ apartments per day. Empty electrical conduits were installed in the slabs and panels at the factory. Plumbing and exhaust ductwork were installed conventionally at the site.

want to assure himself of the structural integrity of the system with respect to the context of its use; conformance with applicable codes; wind resistance of the system.

Consulting mechanical and electrical consulting engineers have to do more than they would normally, the first time the system is used. For example, they have to find out where openings are permitted in the structure (floor and wall panels), and then these have to be accurately laid out.

With prestressed planks, for example, openings cannot cut across prestressing strands. Perhaps, where openings must be large, the planks may need to be cut short, and the opening framed separately. The electrical engineer may have to accurately lay out conduit runs—a function normally performed by the electrical contractor with conventional construction.

Presently, a large portion of the mechanical work is done in the field—installation of piping and ductwork; running of wiring and connection of wires to utilization devices. Probably a certain amount of this will always be done in the field. Piping "walls" can be, and actually have been, prefabricated. Drainage, waste, and vent (DWV) piping and branch water piping lend themselves to the prefabricated approach. But making connections with water risers, because of changing pipe sizes and the type of joint necessary, is more susceptible to misalignment problems, and, with present technology, is better done in the field.

Once the system has been adapted, its use can be repeated—provided that the client and sponsoring authority (if there is one) accept the original design, apartment layouts, utilities, etc. When changes are made, however, then new investigations are necessary — with engineers working out new details and coordinating various aspects of the sub-systems. New shop drawings must be made, and these reviewed by the consulting engineers.

The economy of industrialized building is, of course, adversely affected when a factory has to produce a lot of special panels and other elements. The factory production line has to run smoothly, with "specials" being produced at a rate that integrates with total production so that the total process is continuous. A factory must produce a given minimum number of housing units per year to be profitable. If, for some reason, a project has a lot of specials, then consideration should be given to onsite, rather than factory precasting. Obviously, the more that architects and consulting engineers understand the possibilities and the constraints of a given system—and the more they design with these in mind—the better the cost picture will be.

Yonkers project serves as a "pilot plant" for trying out a French system
Of the more than 50 industrialized building systems in use in Western Europe, Tracoba is one of the largest, with over 80,000 hous-

The structural engineer saved reinforcing steel by taking a sophisticated approach to the design of the floor slab

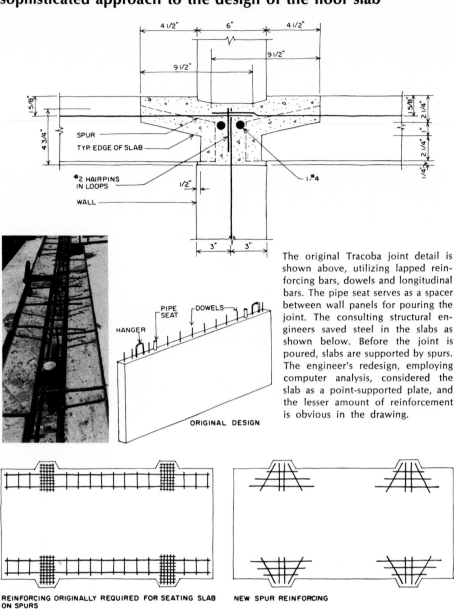

The original Tracoba joint detail is shown above, utilizing lapped reinforcing bars, dowels and longitudinal bars. The pipe seat serves as a spacer between wall panels for pouring the joint. The consulting structural engineers saved steel in the slabs as shown below. Before the joint is poured, slabs are supported by spurs. The engineer's redesign, employing computer analysis, considered the slab as a point-supported plate, and the lesser amount of reinforcement is obvious in the drawing.

REINFORCING ORIGINALLY REQUIRED FOR SEATING SLAB ON SPURS

NEW SPUR REINFORCING

Changes were made in the joint details to give a mechanical connection that would meet recently adopted HUD-FHA criteria

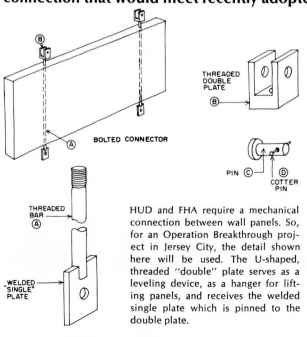

HUD and FHA require a mechanical connection between wall panels. So, for an Operation Breakthrough project in Jersey City, the detail shown here will be used. The U-shaped, threaded "double" plate serves as a leveling device, as a hanger for lifting panels, and receives the welded single plate which is pinned to the double plate.

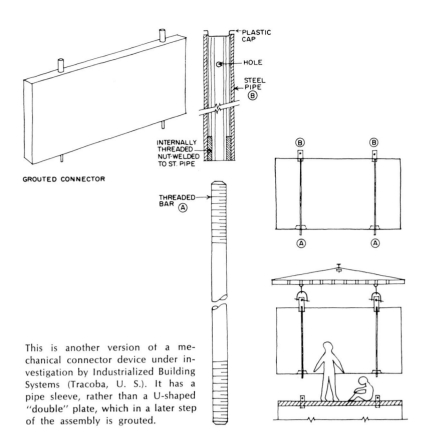

GROUTED CONNECTOR

PLASTIC CAP

HOLE

STEEL PIPE ⓑ

INTERNALLY THREADED NUT-WELDED TO ST. PIPE

THREADED BAR Ⓐ

This is another version of a mechanical connector device under investigation by Industrialized Building Systems (Tracoba, U. S.). It has a pipe sleeve, rather than a U-shaped "double" plate, which in a later step of the assembly is grouted.

Plumbing and ducting were conventional, but required careful coordination with structure. The electrical system utilized new devices; was laid out by the consulting engineer for factory conduit installation

Connections from conduits in one floor slab to conduits in an adjacent one, to handle circuitry common to two rooms, is made with water-tight flexible conduit. The same connector is used to join a conduit in a slab to a conduit in a wall panel. The feed is down from the top of a wall panel to outlets and switches. Positioner boxes placed in the forms fix the location of the conduit terminations. A different device, called a contramold, will be used in the future for more accurate installation of conduits. The water-tight flexible conduit has a limited bending radius which can work within the tolerances of construction, but can be affected if misalignments occur in the factory fabrication of panels.

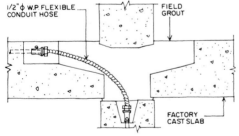

1/2"φ W.P. FLEXIBLE CONDUIT HOSE

FIELD GROUT

FACTORY CAST SLAB

TYPICAL SLAB TO WALL CONNECTION

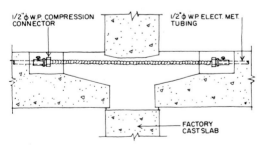

1/2"φ W.P. COMPRESSION CONNECTOR

1/2"φ W.P. ELECT. MET. TUBING

FACTORY CAST SLAB

TYPICAL SLAB TO SLAB CONNECTION

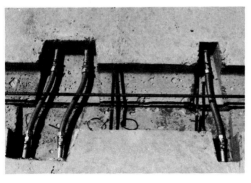

ing units having been built in Europe and North Africa. The system has been adapted to American practice by Industrialized Building Systems, Inc. (IBS) of New York City, and the system was chosen by Module Communities, Inc. (MCI), a building systems developer of Yonkers, New York, as their entry to the market.

Working with IBS on the structural adaptation was Paul Weidlinger, consulting engineer; on mechanical adaption (HVAC and plumbing) it was Cosentini Associates, consulting engineers; and on electrical adaptation it was Eitingon & Schlossberg, consulting engineers who were associated with the Cosentini firm. Architects were Renato Severino and Herbert Rothman.

The structure for the Yonkers project is basically the same as that used in Europe with one important exception. Floor slabs of the system have four "spurs" which support the slabs on the walls until the joint is poured between two contiguous floor panels and the wall slabs above and below them (see structural detail). Such arrangement makes the leveling process simpler. In the European design, the edge of the floor slab abutting a wall was considered as a beam, and was reinforced accordingly. Weidlinger, on the other hand, assumed the slab to perform as a point-supported plate, and redesigned the reinforcement. According to Weidlinger partner Matthys Levy, this saved about ½ lb of reinforcing steel per sq ft of slab. In the alternate design, the slab was analyzed using a fine mesh grid and a computer to obtain the stress pattern. From these results, a reinforcement pattern radiating from the spurs was proposed. A prototype panel tested with the new spur reinforcing substantiated the safety of the design.

Another structural modification—a different joint connection—was developed for Tracoba projects in which there is FHA and/or HUD involvement. The typical Tracoba joint provides lapped reinforcing bars in the joint between floor slabs and wall slabs. But FHA and HUD criteria, adopted for Operation Breakthrough, call for a more conservative design. These criteria follow the design criteria for joining panel structures that were developed in England after an investigation into the collapse that occurred at Ronan Point after a gas explosion in a corner apartment set off a chain of collapse of all of the 18 floors below it. FHA and HUD criteria specify welded or bolted joints, only. The new Tracoba detail developed to conform to the criteria, shown on page 60. This connector was used in Operation Breakthrough projects in Jersey City and for three apartment towers in the Twin-Parks project in the Bronx sponsored by the Urban Development Corporation.

A different connector device utilizing a grouted sleeve in lieu of a mechanical connector is presently being tested. This device costs somewhat less and will be adopted once field tests are finished.

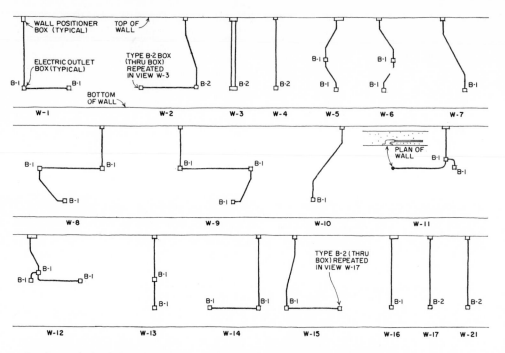

Electrical wiring had to be worked out in detail by the consulting engineers

The Yonkers project was not designed with the idea that it would serve as a prototype for successive projects. Rather it was a project with which the building systems developer could, so to speak, "cut his eye teeth." Apartments are spacious and will rent at the higher end of the scale. The exterior, happily, is attractively designed.

Because the Tracoba system uses short spans for floor slabs, almost all walls are bearing walls. This means that a lot of the wiring is buried in the wall slabs as well as in the floor slabs. Some dry walls are used to conceal the main electrical risers and to conceal plumbing.

There are more electrical risers in the Yonkers project than would normally be encountered: apartments are individually metered, and all meters are located in the basement, necessitating more feeders than if the building were centrally metered. Apartments will have electric baseboard heating and through-the-wall air conditioners.

In Europe, Tracoba uses plastic conduit for the buried wiring. In the Yonkers project thin-wall conduit (EMT) had to be used. Connections from one floor slab, across a joint, to another floor slab, and connections from floor slab to wall slab were made with flexible, water-tight connectors that were approved for this application by Underwriters' Laboratories. These connectors can accommodate to the usual tolerance expected with panelized construction. Positioner boxes which establish terminal points for conduits in the concrete panels have to be carefully installed in the slab in the factory, however, because the flexible connectors have a limited bending radius. The conduit in a floor slab has to line up fairly closely with the conduit in the wall panel. A device called a contramold is being used in the slabs for succeeding projects to assure accurate spacings of conduits—with slabs installed within tolerances, conduits should line up.

In the Yonkers project, many different wall conduit situations were necessary. These were identified and keyed to an electrical plan (such as an electrical contractor ordinarily would do) by the consulting electrical engineer (see drawings this page). The engineer showed the exact conduit shapes and locations for both wall and floor slabs.

The empty conduit was installed in the MCI factory, located in the Bronx, by laborers supervised by a New York City (Local 3) union electrician. The electrical contractor pulled the wires in the field and made all necessary connections to utilization devices (switches, lights, appliances, convenience outlets, etc.). An electrician had to be on hand during the erection of the precast concrete panels to install the sections of flexible water-tight connectors, inasmuch as they are contained within the poured concrete joint between panels.

The electrical plan below shows how the consulting engineer laid out exactly the conduit shapes and locations so that the conduit could be cut, bent and installed in panels at the factory. Positioner boxes are shown in adjacent floor slabs where conduits must be joined by the flexible connectors. Also identified on this plan (by numbers in the hexagons) are the various wall panel conduit conditions—nearly 40 in all—that occur throughout the building. The wall panel conditions in the area of the floor plan given are shown in the drawings above. Electrical boxes are indicated.

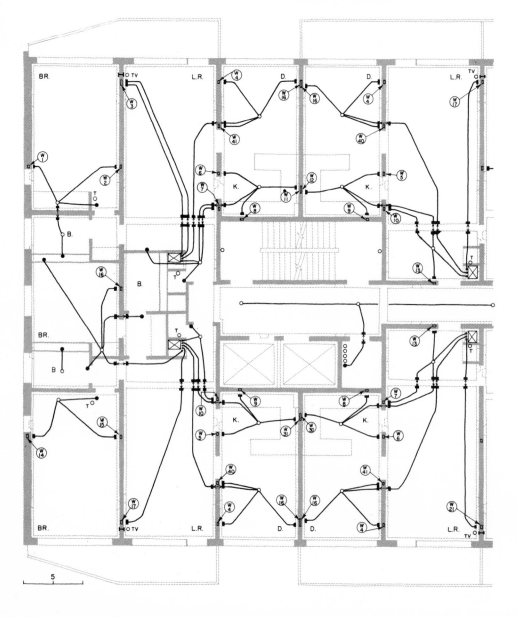

High-rise concrete housing system for dormitories

Jim Dallas photos

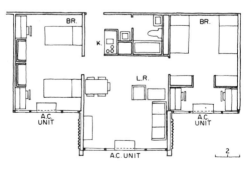

On the University of Delaware's Newark campus are two high-rise dormitory buildings constructed with the Bison factory precast concrete system that has been highly successful in England. The system used for the dormitories, designed by Charles Luckman Associates, has precast, load-bearing exterior and interior walls, 27-ft prestressed concrete planks, and an aluminum and glass infill between precast spandrels.

Several European industrialized housing systems, of which the Bison system is one, have been franchised in the United States and Canada. While only a few projects, using several of the systems, have been completed so far, a modicum of experience has accumulated, and, importantly, the disciplines of the industrialized housing process, based upon the concept of factory-produced structural components, are beginning to be understood. Further, the professionals who have worked with these systems are getting an idea of what these systems can and cannot do.

Proposed originally in steel, the structure was switched to the precast system

University of Delaware housing officials decided to take the private developer route in getting their dormitories built, and sponsored a competition whose entries were judged on the basis of quality of architectural concept and cost. The winning entry was that of Ogden Development Corporation, headed by Charles Luckman, in a joint venture with Frederic G. Krapf and Son, Inc., Wilmington general contractor. In 1971, a $10.5-million contract was let for the two dormitory towers and a 27,000 sq ft student commons, the project being designed to accommodate 1,300 students. The towers have 375,000 sq ft and incorporate 255 one-bedroom apartment units and 197 two-bedroom units.

The Ogden Development Corporation-Krapf joint-venture's original proposal was for a conventional steel-frame design. Shortly after winning the contract they learned that they could obtain the industrialized concrete system without an increase in cost, while at the same time

The building shell uses bearing walls and prestressed slabs. The 8-ft-wide slabs are cast and pretensioned in a continuous bed and cut apart after the concrete sets. Exterior wall panels are faced with white architectural concrete in a fluted pattern, outlined by smooth spandrels and corners. They are of sandwich construction with a core of foamed polystyrene insulation, and an inner layer of ordinary concrete. Through joints between wall panels are protected from the weather by the rain-screen technique—a baffle set in grooves of adjacent panels keeps out rain while equalizing pressure. Panels and planks were erected using a tower crane with maximum weight of panels being 10 tons.

gaining some square footage in the apartments. The proposal for supplying the system's concrete units was made by Strescon Industries, Inc. of Baltimore. Further, the system promised improved acoustical privacy, interior finishing and maintenance.

The same basic floor plans as originally worked out were retained, with the exception that one-bedroom and two-bedroom apartments were grouped so that bearing walls would align across the short dimension of the plan, a condition preferred by the structural engineers for shearwall design. Also, the depth of the floor plan was adjusted to match the 8-ft-width module of the floor planks.

Because the floor plans were changed only to this extent, the architect found that a much larger variety of wall panels was required than would have been the case if the concrete system had been selected at the start, and the floor plans laid out considering the nature of the system. The variations consisted mainly of different types of panel connection details, different reinforcing patterns, slight differences in dimensions, etc. Over half of the panels on a typical floor had some variation, even though minor. But the original plans were retained because redesign would have cost both time and money.

The collaborative efforts of those participating in the project have paid off in terms of high-quality appearance as well as in construction time—the structure was erected at a rate of one floor per week per building. This meant that the plumbing and electrical trades were inside for their work—which was done conventionally on site—much sooner.

For the architect, the Luckman firm sees a reduction in the number of working drawings required. Of course he still must prepare the floor plans; perhaps detail an infill curtain wall, and do normal interior detailing for bathrooms, kitchens, door bucks, etc.

But, the architect and structural engineer found—as others have—that the checking of shop drawings on a building that has not been done before takes considerable time. Of course, if the same sys-

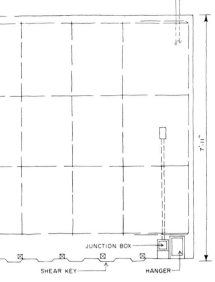

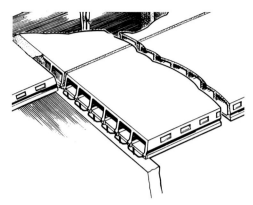

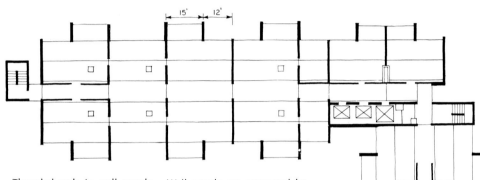

JUNCTION BOX

SHEAR KEY——— ———HANGER

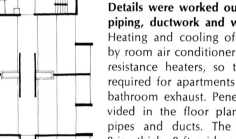

Threaded rods in wall panels serve two main purposes: 1) they are used for leveling of the panels; 2) they take tensile loads created by wind forces in some panels. The hanger box provides continuity from a rod in one panel to those in panels above and below. The hanger boxes also serve as means for leveling. Bottoms of panels have shear keys to transmit forces to the floor diaphragm. Electrical boxes are cast in the panels as seen above. After panels are leveled, drypack concrete is put underneath.

Wall panels are supported by spider braces until planks are set, and corner joints poured.

Bearing walls support floor planks which are 27-ft long except for projected areas, where they are 15 ft. Planks at corridors use "stretched-out" point supports at the corners. These panels have a solid ring of concrete around the perimeter to minimize deflection and to transmit loads in shear.

Openings in the floor slabs for plumbing and ducts sometimes required special support and/or reinforcement.

tem is used again by the same architect and engineer, they will have familiarity with various details. This familiarity, however, will not be of much help if different industrialized systems, with different details, are used on succeeding jobs.

Details were worked out to accommodate piping, ductwork and wiring

Heating and cooling of the apartments is by room air conditioners that have electric resistance heaters, so the only ductwork required for apartments is for kitchen and bathroom exhaust. Penetrations were provided in the floor planks for passage of pipes and ducts. The prestressed slabs, 8-in. thick, 8-ft wide and 27-ft (or 15-ft) long, are hollow-core, ribbed units. Some openings were provided by putting blockouts in the continuous forms. In other cases, they were cut out after the concrete had set and the slabs cut to length.

For small penetrations needed for the plumbing wall, the structural engineer permitted a series of openings across the width of the slab made by cutting out top and bottom surfaces, but preserving the ribs intact. In some cases openings were made by stopping a slab short of a bearing wall, the slab being supported by a steel collar. In such cases the slab was stiffened at the end by chopping out the top part of the slab and filling the void with concrete. Where large openings were required, the engineer allowed a maximum of two ribs to be cut (see drawing, page 66). Additional shear reinforcement was provided in the area where the opening was to be cut so that load would be transferred to the other ribs.

No wiring is run within the wall panels or the floor slabs. Because of the long span of the slabs, and the need for only occasional shear walls along the corridors, many of the partitions could be dry wall, with wiring being run within these. Where outlets were needed in bearing walls, the wire was run in a recess at the bottom of the walls made as the drypack under the walls was tamped. The recess was covered by a metal plate held by clips fastened to wooden plugs cast in the panels.

The drawing shows the types of openings provided in the floor planks for penetration of ducts, plumbing, and electrical risers. Planks with the large opening had to have additional reinforcement in that area. Room air conditioners with electric heating elements are used to maintain thermal comfort—thus, the louvered area for the air-cooled condenser. Because of the long spans and design of corridor planks, considerable lengths of dry-wall partitions were possible, making it easy to run flexible electrical cable.

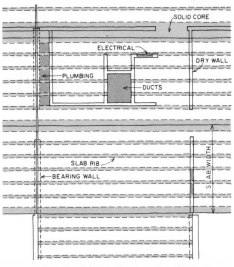

Wind resistance had to be thought out carefully to minimize stresses and costs
Shear wall action had to be depended upon to resist wind loads because it was not possible to create a moment-resisting frame by tying bearing walls together across corridors with dropped beams—a condition the architect wanted to avoid. (The connection could not be worked out within the shallow 8-in. depth of the slabs.) In any event, it would have been difficult and expensive to develop moment resistance. Wind stress analysis was made by the engineers—Severud, Perrone, Sturm, Bandel—using a computer program.

Because of the L-shaped plan, the shear walls had to be carefully located to avoid an eccentric condition with respect to the center of "stiffness" of the building. Eccentricity would have greatly increased the wind moment which would have overstressed the shear walls. The structural engineer avoided this condition by judiciously placing shear walls along the corridors, and by utilizing a long shear wall at the elevator core.

The wall panels are connected to each other in the vertical direction by means of 1-in. diameter rods. Depending upon the dead load, and the particular location of the panel in the building, the panels may be put in tension by wind load, or they may be always in compression. When tension forces may occur, the rods are continuous from top to bottom, being anchored to the foundation, and tied from one panel to another by means of steel hanger boxes set in the panels, with nuts being turned down on the threaded rods to secure them. When a compressive condition exists, steel angles—which cost less than the hanger boxes—come attached to the wall panels, and the rods are needed only for erection leveling and stability.

RESIDENCE HALL HOUSING, University of Delaware. Architects: *Charles Luckman Associates;* structural engineers: *Severud Associates;* mechanical engineers: *Cosentini Associates;* electrical engineers: *Eitingon & Schlossberg Associates;* joint-venture developer: *Ogden Development Corporation and Frederic G. Krapf and Son, Inc.*

Long-span precast system for a science center

*Harvard University Science Center,
Cambridge, Massachusetts
Architects: Sert, Jackson and Associates
Consulting structural engineers:
Lev Zetlin Associates, Inc.
Consulting mechanical and electrical engineers:
Syska & Hennessy, Inc.
Cost consultant and construction management:
Turner Construction Company*

When, after years of fund raising, Harvard University received a $12.5 million gift in June 1968 toward their projected Science Center, the administration wanted to move ahead quickly. Construction of the $17.7 million facility began in the summer of 1970, and was completed in about 30 months. Contributing to the speed of construction was the precast concrete framing. Structure for the 400- by 60-foot laboratory wing was erected in seven months. Precast concrete is used everywhere except for the structure—which is steel—that houses four demonstration theaters, and the special basins and supporting structures—which are in-situ concrete—for the three roof-top cooling towers. These cooling towers serve a below-grade chilled-water plant that provides air conditioning not only for the Science Center, but for all of the Harvard campus.

Besides the laboratory wing, the complex includes a theater building, a large stepped-down structure for the mathematics department, a science library, and a low portion for administrative offices.

The University signed a negotiated contract with Turner Construction Company for cost-consulting and construction-management services. Based upon nearly complete working drawings, Turner's estimate was that the building would cost $19 million; the University's budget was $17 million. A number of things were done to cut costs, and among them were: 1) structure and mechanicals are left exposed in the laboratory wing; 2) concrete fill over the precast decking will serve as the finished floor in the laboratory wing; 3) partitions are dry wall rather than block in the laboratory wing; 4) ductwork for fume-hood exhaust is galvanized with an interior coating, rather than stainless steel; 5) air-conditioning zoning was cut to a minimum (for example, there is no zoning for exposure); 6) two-pipe fan-coil units are used for classrooms; 7) clear glass is used rather than tinted (operable sash is provided in the laboratory and classroom wings).

The faculty wanted clear space in the laboratory wing, even though there were to be offices along the corridor that is located along the south side of the building. Early on, a poured-in-place framing scheme was considered, utilizing precast decking. Three rows of columns were con-

Bennett Jones

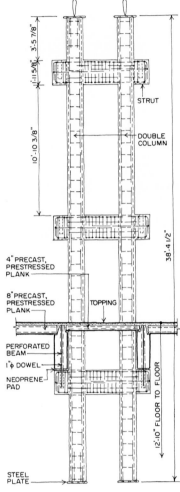

STRUT

DOUBLE COLUMN

3'-5 7/8"

1'-11 5/8"

10'-10 3/8"

38'-4 1/2"

4" PRECAST, PRESTRESSED PLANK

8" PRECAST, PRESTRESSED PLANK

TOPPING

PERFORATED BEAM

1" φ DOWEL

NEOPRENE PAD

12'-10" FLOOR TO FLOOR

STEEL PLATE

The fully precast frame of the laboratory wing is comprised of "ladder" columns (in three-story-high units) on one side of the 60-ft deep building and U-shaped post-tensioned mechanical shafts on the opposite side—spanned by perforated concrete girders which support prestressed, hollow-core decking. The columnar elements are on 24-ft centers. The "ladders" are used two-high, and are connected by threaded rods and leveling nuts as shown below. Wind forces are taken by the post-tensioned shafts which have six 1⅝-in. rods in the outer face. The shaft units were cast in three separate pieces and joined as shown in the detail. Faculty wanted a 9-ft depth from floor to under side of beams; with the girder being 3-ft deep to the offset, the prestressed decking, 8 in., and the topping, 2 in., floor-to-floor height is 12 ft 10 in.

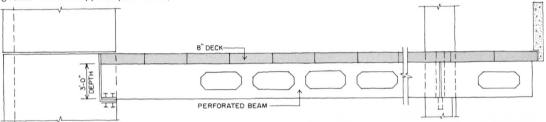

8" DECK

3'-0" DEPTH

PERFORATED BEAM

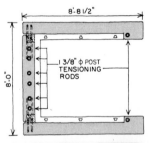

8'-8 1/2"

8'-0"

1 3/8" φ POST TENSIONING RODS

templated. The corridor was cantilevered 10 ft; the next row of columns came 15 ft away, along the office partitions; and they were followed by 35 ft of clear lab space to the third row of columns.

But many of the columns came at awkward locations, and the faculty argued that the presence of the interior columns inhibited flexibility. Turner was concerned about uncertainties with respect to the cost and construction time for poured-in-place concrete. Furthermore, the Boston area is a favorable market for precast concrete as there are several local companies.

Pairs of long perforated girders give a 50-ft clear span for the laboratories

The structural solution to the laboratory wing, developed by engineer Lev Zetlin, was a totally precast system with 3 ft-8 in. deep, 60-ft long reinforced concrete girders—perforated for utilities—spanning between "ladder" columns on one side of the building and post-tensioned, U-shaped concrete shafts on the other. The girders, which cantilever 7 ft over the single corridor, are used in pairs (7 ft apart) on 24-ft centers. The 24-ft module is a convenient one for the laboratory layout (fume hoods are generally on 12-ft centers).

The heavily-reinforced concrete girders have a maximum of penetrations—considering what could be permitted structurally. The penetrations are not always utilized, but for economic reasons, all girders were made the same.

The "ladder" columns were cast in units three stories high, and have haunches to carry the perforated girders. The girders are "simply" supported on the haunches, being dowel connected. The time-saving arrangement used for joining an upper "ladder" column to the lower one involved using steel plates, threaded rods and leveling nuts, as can be seen in the photo and the drawing across page.

Wind loads on the laboratory wing are taken entirely by the post-tensioned concrete shafts, themselves; no rigid-frame action has been employed. Post-tensioning force was designed to prevent any tension

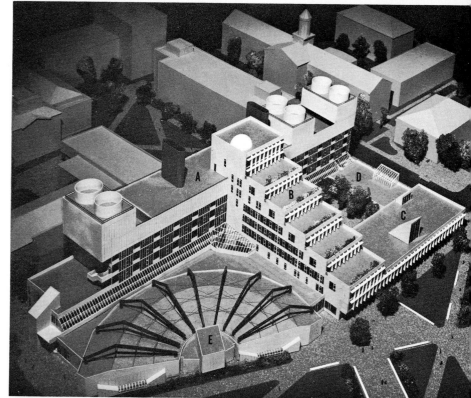

The Science Center is comprised of a laboratory wing (A), a classroom wing for the mathematics department (B), a science library (C), administrative wing (D), and a structure housing four demonstration theaters (E). Atop the five-story lab wing are structures to house three cooling towers. At left is "simply-supported" precast frame used for the math wing. Wind resistance is provided at elevator and mechanical shafts.

from occuring in the outer faces of the shafts when loaded by the "design" wind.

Arrangement of supply chases means some long duct runs in the laboratory wing

Given the choice, mechanical engineers would prefer that the ceiling space in complicated buildings such as laboratories and hospitals be clear of structural impediments. It is difficult—and sometimes impossible—to organize the myriad utilities in coherent, repeatable, modular arrangements. One approach is to have a large number of vertical shafts and short horizontal runs serving each laboratory module. When the plan precludes this, then the engineer hopes to have as many large openings as possible in beams and girders—and even then, tight conditions may occur.

The Science Center laboratory wing has 15 mechanical shafts, 24 ft apart, with 6- by 8-ft interior dimensions, running from the basement to the fifth (mechanical) floor. Mainly they are used for ducts handling exhaust from fume hoods, (12 ft apart) but they also carry some plumbing and electrical risers. On the opposite side of the building, 6- by 12-ft chases for supply air ducts are interspersed between office areas and are spaced 48 ft, 72 ft, 96 ft and 120 ft apart, which obviously results in some long horizontal duct runs, some fairly large duct sizes, and some fairly large reheat coils to be fitted in. There are also small chases at each of the "ladder" columns for piping. In the chemistry laboratory areas supply air is 100 per cent fresh air, and all exhaust is taken through the fume hoods.

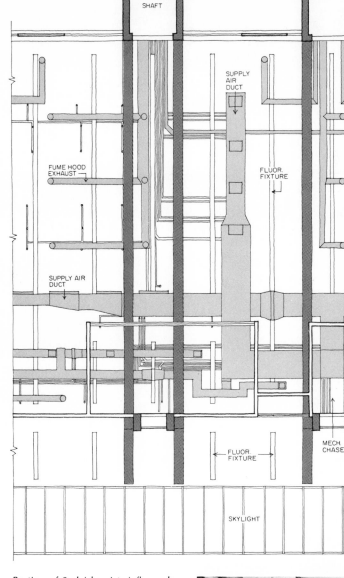

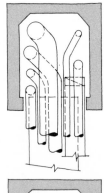
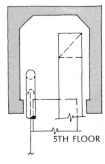

5TH FLOOR

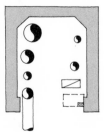
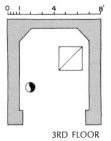

3RD FLOOR

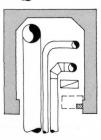
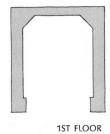

1ST FLOOR

Portion of 2nd (chemistry) floor plan shows typical supply-air ducting and fume-hood exhaust ducting. Exhaust ducts go up concrete shafts to 5th floor where in-line fans expel fumes to outdoors. Ten of the 15 shafts are heavily utilized (example, far left), and five, only lightly (example, near left). Photos: girder in basement (right, top); girder in chemistry lab (right, bottom); view toward mechanical shaft (below).

Prefabricated concrete boxes for hotel construction

The secret to getting the 500-room 21-story Hilton Palacio Del Rio Hotel completed in time for San Antonio's Hemis-Fair (April 1969) was the off-site construction and fitting-out of concrete boxes comprising the hotel guest rooms. The builder, H. B. Zachry, credits this technique with cutting construction time by one-third. The guest units, which begin on the fifth floor, were fabricated in a $500,000 casting yard located seven miles from the site. At the site the units were hoisted into place completely equipped with bathroom fixtures, air-conditioning fan-coil unit, wiring, glazing—and even the furniture. Costs of the concrete boxes for this $7.5-million hotel were said to run about $10 per sq ft of floor area (1968).

Another reason for prefabricating most of the building at a remote location was limited access to the 350- by 40-ft site. Only two blocks from the Alamo, the site was bounded on the east by a road already choked with heavy equipment and trucks rushing HemisFair construction, on the west by the Paseo del Rio

(River Walk) with its constant stream of sightseeing boats and pedestrians, and on the north by one of the city's main thoroughfares. On the south, jammed up against the construction site, was a recently restored 120-year-old building which could not be disturbed. This left little room for maneuvering equipment.

In spring 1967, officials of the city of San Antonio and of HemisFair (of which Zachry was board chairman) became alarmed over the fact that the city lacked sufficient hotel space to accommodate fair visitors and to support the $10.5-million convention center.

When it became apparent that no one else would attempt the project, Zachry acquired the site and in July started to work on the foundation. Structural engineers had told him his concept of the highrise hotel built of precast boxes was feasible, but he had no final architectural plans and had to get city council approval to begin construction without a building permit. (Plans were completed, and the permit issued, three months after construction

began.)

Project architects Cerna and Garza, of San Antonio, laid out the guest-room floor plan around the structural requirements of the boxes. They were already familiar with the module, however, since Zachry had been working with them on possible uses for it. (Earlier, as an experiment, Zachry had used 108 of the boxes in constructing a two-story motel on Padre Island off the Texas Gulf Coast. Structural design of these boxes was entirely different from that of the hotel modules, however.)

The hotel has 19 stories on the street side, 21 on the river side. The lower levels will house shops, restaurants and meeting rooms. While these were being built conventionally, and the elevator-service core was being slipformed, Zachry was setting up the casting yard at his headquarters. The first room was hoisted in November 1967. Originally, the schedule was to set 10 rooms a day but a special hoisting device speeded this to a maximum of 22 boxes in one day and an average of

Two assembly lines were set up at the $500,000 casting yard to build the concrete boxes which comprise all guest-room units of the $7½-million Hilton Palacio del Rio. Each line has eight sets of forms, and total production at the yard was eight units per day. Rail-traveling gantries hoisted boxes from the forms after the concrete had preset for three hours and had been steam-cured for seven hours. The boxes were then trucked to the finishing yard. Steel reinforcement was mainly No. 5 bars spaced about 6 in. o.c. Wire rope strands tied into the reinforcement were looped to form "handles" for lifting boxes by gantry and crane.

Two sizes of boxes were used, giving two room sizes as well as a variation in texture of the facade. A 20-in. space was left between units to provide chases for installing plumbing and air-conditioning piping. Ceiling-hung fan-coil units were used, supplied by a four-pipe system.

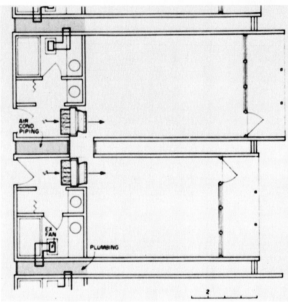

between 16 and 18 per day, even on the top floors.

Each guest-room floor—from floor five to 19—has 31 modules. The 20th floor has four three-room suites. Boxes which made up these suites were cast with certain walls which could be eliminated to create the larger facilities. The 21st floor, which houses a banquet hall and other dining rooms and clubs, was conventionally built.

Operations of the off-site casting yard were carefully scheduled

Casting yard crews worked staggered shifts on a 24-hour basis, five or six days a week. First step in casting the units was to pivot and bolt the permanent, hinged outer forms into place. Eight reinforcing crews—one for each pair of modules—worked from 10 p.m. to 5 a.m., then four additional crews came on at 5 o'clock to finish the steelwork. Along with the steel, wiring conduit and electrical boxes were installed. Screeds were put in ready to form the floor and the 4-in.-high curb.

There were about two tons of steel in each unit, mostly No. 5 bars spaced approximately 6 in. o. c. (The amount of steel varied in floors, walls and ceilings; it also varied depending on location and the load the unit carried.) Wire rope strands tied into the rebar framing at the top corners of each unit were looped up to provide permanently cast "handles" for hoisting on and off site by the gantries and crane.

Steel, like everything else on the job, was prefabbed. Three men, working in a central reinforcing bar yard prepared the steel. Before the steel crews left the job, their work was inspected by three quality control men.

Each module had its permanent room number assigned to it when it was poured in order to have it ready for placing in the proper sequence and to determine the location of door frames and mechanical and electrical outlets. Frames were cast right in the concrete. Foamed plastic blockouts, most of them inserted from the outside of the form in order to

simplify stripping, were used for mechanical connections.

Concrete placing started at 6 a.m. (because of the additional cost of working earlier than 5 a.m. or later than 8 p.m.). Crews consisted of seven concrete men, plus a crane operator and oiler. At 10 a.m. two additional men came on to trowel off the floors and give them a hand finish.

The 16 permanent forms for casting the units were arranged in two rows of eight each, with rails alongside these production lines so the custom-designed gantries could move freely to lift the finished rooms from the forms.

Crews worked on modules in pairs, using turntables—made of 12-in. WP beams built up and rolled and mounted on large dolly wheels—to transfer the inner form from a finished module to its opposite number.

For jacking the inner forms into place and out, the contractor simply used the floor of the box as a base. For future jobs, hydraulic jacks with preset stops

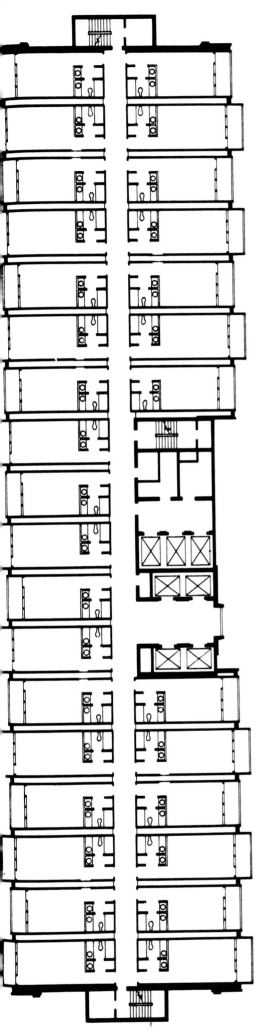

At the finishing yard a rubber-tired gantry took the box from a truck and set it on blocks, or put it on top of another box. All interior work was performed here including partitioning, setting of the plumbing fixtures and installation of the fan-coil unit. At the site plumbing and air-conditioning piping connections were made to risers in the chases. Glass and balcony railing were also installed at the finishing yard. Units were carpeted and the furniture was put in at this stage.

To keep the boxes from swaying in the wind, the builder devised a stabilizer consisting of a helicopter rotor. The vertical steel section is a weld plate which holds in place the brick-veneered precast concrete closure panel. The reinforcing rods seen extending from the weld plate were welded to those of an adjacent box behind the brick closure panel.

will be used to improve the operation.

In order to give the facade of the hotel a large-scale texture, modules were made in two sizes. Outside dimensions were 13 ft wide, 9 ft, 1 in. high and either 32 ft, 8 in. or 29 ft, 8 in. long. Floors and walls are 5 in. thick, while ceilings are 4 in. thick.

Pour for a floor took 20 minutes while the walls and ceiling took half an hour. As soon as concrete men made the final pour on a box, utility crews moved in and covered it with a tarpaulin. Concrete was allowed to preset for three hours, then was steam-cured for seven hours with steam at 40 lb pressure and 150 F.

Finished boxes were picked up by a railed gantry and hoisted onto a truck to be transported to the finishing yard, where they were handled by a rubber-tired gantry.

When the truck delivered a box to the finishing yard, the rubber-tired gantry lifted it onto blocks on the ground or stacked it on top of another box (at one time most of the rows were double-decked to conserve

storage space).

First operation in the finishing yard was to rough in the dry walls and install the bathtub and ceiling-mounted air-handling unit (the hotel has a four-pipe system and will be supplied with hot and chilled water by a central plant serving the entire HemisFair). Wiring and plumbing was completed, walls finished with a vinyl covering, floors carpeted and the room completely furnished and cleaned, ready for company. Then the room was sealed for the move downtown.

Hoisting boxes into place on-site
was a tricky problem, ingeniously solved
One problem imposed by the limited site was the impossibility of using taglines to guide the boxes on the 200-ft lift. The contractor anticipated trouble in keeping the rooms steady, level and properly oriented, particularly on windy days.

The ingenious answer to this problem was a stabilizing device, consisting of a pipe frame which has a tail rotor and

drive from a Sikorsky helicopter at one end and the engine at the other, keeping the weight balanced.

Originally no one was permitted to ride a box up, so the stabilizer was equipped with an airplane autopilot to control the rotor to a specified heading until the box reached the top. The rotor could also be controlled from the ground with a manual pushbutton control.

However, it became so easy to handle the boxes that men started riding them up so the job of unhooking them could be speeded and made more convenient. An operator steered the box with a manual control powered by the motor from an automobile window lift. (The autopilot will be used on future jobs where manual control is not feasible.) A rigger rode with him to help with the unhooking.

With the stabilizer, crews were able to set a box in less than 20 minutes from hookup time. Because box placement had to match the elevator shaft, each unit had to be set exactly on the box underneath and at a precise elevation. The con-

tractor was working to a tolerance of ¾ in. and had to prevent creep.

Final step: tying the precast units into a unified structure

Boxes were placed 20 in. apart to provide a mechanical chase. Plumbing and wiring were run up the chase for quick connection to individual rooms (no plumbing or wiring goes through the floors of the units). Removable panels in the hotel corridor provide access to the chase for installation and maintenance work. The floorless chase is crossed every 3 to 4 ft by steel plates to provide a base for a temporary working floor between rooms. The chase is closed on the exterior by a continuous brick panel.

For leveling, boxes were placed onto steel shim plates and on four 5- by 8-in. grout pads (averaging ¾-in. thick) which were made of polyester concrete. The concrete sets up in about 30 min and is strong enough to hold a box in three hours. Between the pads crews put a continuous rope of grout around the perimeter

of the box to give an airtight seal and continuous contact. Additional connections were provided by weldments around the bottom and vertically.

The rear of each box had a 2½-ft haunched cantilevered shelf with reinforcing bars extending out an additional foot. To make the corridors, these exposed bars were laced together and a concrete key strip was poured to connect the boxes and create a corridor floor-ceiling.

Half of the room units—the long ones—have a 4-ft slab for the roof of the balcony, with approximately 1½-ft open space being left between slab and box. When the short box is placed above the long box, about a 2-in.-wide space is left which provides a key joint for tying the boxes together.

The mechanical chases at the exterior were closed in by precast concrete strips to which prefabricated brick panels had been attached, all of this being done at the casting yard.

The precast slabs were attached to the boxes with weld plates and anchors and became part of the structure. Metal brick anchors were cast into these slabs for attaching the decorative brick panel facing.

Interior closures for the mechanical chase were removable panels which provide access for maintenance. Concrete for the hotel modules was a 7½-sack mix with lightweight aggregate and a 4-in. slump which developed a strength of 5,000 psi.

Owner of the hotel is Palacio Del Rio, Inc. James D. Lang is vice-president of Palacio Del Rio, Inc., in charge of engineering and architecture. Larry J. Raba is architectural coordinator.

Architects are Cerna, Garza & Associates. Feigenspan & Pinnell are structural engineers. Schuchart & Associates are mechanical-electrical engineers.

Flying forms for concrete construction

Irregular tower facades and column spacings were a challenge for the technique, but it came out cheaper

Maximum economies in the application of flying forms to multistory concrete construction call for a building layout that allows large-size, consistently-shaped forms. But even with the irregularly-shaped facade and varied column spacing of the 21- and 9-story structures for New Hope Towers in Stamford, Connecticut, designed by architect Robert L. Wilson, the contractor reported significant savings with the method. Beyond the economies made possible by the flying forms, additional cost savings were achieved by precasting the wall panels at the site. To facilitate erection of the panels to the boxed-in recesses of the facade, a counter-balanced boom was used.

On New Hope Towers, the contractor, Frank Mercede and Sons, Incorporated used 14 table forms—ranging in width from 14 to 20 ft for the three different column spacings and in depth a maximum of 30 to 45 ft—to cast each floor. Rate of construction for the 6700-sq-ft floors was one floor every four or five days. The flying forms were supported and leveled by jacks attached to the sides of the columns. The forms, being set on the jacks, and having no cross-bracing, allowed freedom of movement for workers. The steel edges of the forms slide out of the building on wheels attached to the jacks. They were inserted and removed from all four sides of the building. Surface of the forms is a special plywood material imported from Finland. Infill panels the width of the columns finished the floor-form surface. Table forms and concrete were lifted by means of climbing cranes.

From a construction economics standpoint (related to crane capacity, labor capability, and form reuse), the floor areas, ideally would have been closer to 9000 sq ft, and both towers 21-stories high. As it was, the tall building had to carry the costs of the low building.

With conventional flat-plate construction, the engineer can locate the columns somewhat irregularly to suit the architectural layout, but in this case, the engineer Viggo Bonneson had to center column spacings so that the flying forms could be used. This discipline, in his opinion, can actually improve building layout rather than hinder it.

Steel-framed flying forms with different widths and depths are hoisted by climbing crane, inserted between columns, and set atop screw jacks attached to the columns. The forms slide on rollers that are part of the jack assemblies. Concrete panels for the modulated exteriors of the 9- and 21-story towers in Stamford, Connecticut were precast at the site.

The speed of cycling the forms is directly related to the repetition and typicality of the structural volumes

Economics of the flying-form technique were put to a hard test in two housing projects of basically similar design in Yonkers, New York, and New York City's Roosevelt (formerly, Welfare) Island.

Because the objective of using flying forms is to speed on-site construction, builders are happiest with simple, box-like structures with double-loaded corridors, and the same floor plan repeated from base to top. But while this approach may keep costs down, it also may inhibit good architecture.

These projects—designed by architects Sert, Jackson & Associates for New York State's Urban Development Corporation—have low-rise wings stepping up to towers as high as 21 stories. Fire stairs and elevator shafts are external. Elevatoring is skipstop, which means two different floor plans, with a single-loaded corridor provided every third floor. Depth of the buildings is 38 ft, except when bays increase it to 41 or 44 ft.

The speed of the flying-form construction process was limited by several factors which will be discussed later. Of course a builder wants construction operations to permit a smooth flow of construction cycles with as little lost time as possible between them. Also, he always looks for ways to cut construction time of the various steps within a cycle.

The Roosevelt Island project (1005 units) was at one time designed for flat-plate construction but the price from Building Systems Housing Corporation was attractive enough for UDC to have the architects and the structural engineers, Weidlinger Associates, change the design to a shear wall structure to accommodate the developer's system of wall forms. Riverview, Phase I, in Yonkers (454 units) was contracted for later, and was shear wall from the start. (Building Systems Housing Corporation is the development arm, and Concrete Building Systems, the construction arm, of Building Systems, Inc. of Cleveland.)

The system also called for the floor slabs to be post-tensioned. Span of the floor slabs is 22 ft, while the post-tensioning cables generally were 90-ft-long, though sometimes 180 ft. The building anticipated that with post-tensioning there might be time-savings in laying the reinforcing steel, and that, hopefully, the lesser amount of steel theoretically possible would save money, but, in the end, codes and other constraints made the costs about the same as for conventional steel reinforcement.

Five different types of forms were used: 1) table forms for the slabs; 2) self-braced steel forms, imported from the Netherlands, for interior walls; 3) custom steel forms for exposed concrete of the stair towers; 4) and 5), custom forms for exposed concrete of elevator towers and for end walls of buildings.

Construction cycling for the wall and floor slabs was the most efficient in the stretches of repeatable elements between stair towers, where construction could be done in a stepped, checkerboard fashion. But construction in the vicinity of the stair towers (called "knuckles" by the architect) was governed by how fast the towers could be erected. Reason was that these towers take the wind load in the longitudinal direction, and the horizontal framing of stair platforms had to be tied to the floor structure of the main building frame. This meant that the post-tensioning cables of the floor slabs had to be a part of the stair platforms. Alignment, fastening, and stripping of the tower forms for the exposed architectural concrete turned out to be time-consuming which, in turn, slowed down construction of the floors and walls.

The construction sequence is as follows, assuming a floor slab is ready for wall forms:

1) wall forms are set; reinforcing steel and conduit are placed; walls are poured (one day);

2) wall forms are stripped; table forms are pulled (rolling on dollies), from another location and reset on a finished floor (one day);

3) edge forms for table forms are set; bottom steel is laid; post-tensioning cables are laid; conduit is installed; top steel is laid; concrete is placed (one to two-and-one-half days);

4) three days later cables are post-tensioned (taking only 2 hours).

The builder would have to be characterized as adventurous to bring this system into the New York City market area where conventional flat plate and flat slab construction have such a foothold. Concrete Building Systems' forming method was not new to them, however, as they had used it previously for box-shaped housing in Cleveland and Boston, and also for a project in Brooklyn.

Bay-size table forms and steel wall forms were sequenced in checkerboard fashion to produce the wallbearing cellular structural frames for two New York State Urban Development Corporation housing projects in the Greater New York area. The designs consist of low-rise wings stepping up to towers. The end walls and the appended stair towers and elevator shafts were produced with custom forms on a floor-by-floor basis. Floor slabs were post-tensioned.

Robert E. Fischer photos

77

A symmetrical free-form tower was a likely candidate for cycling one floor of forms at a time

A forming system that included unusual pie-shaped table forms and combination wall-column steel forms was able to produce one floor a week for this 25-story condominium in Portland, Oregon. All of the forms for a floor were stripped, "flown," and reset in just over three hours. Precast stairs were used to eliminate formwork within the building itself.

Use of flying forms for the 25-story Portland Plaza condominium (by Daniel, Mann, Johnson & Mendenhall, architects and engineers) reduced construction of the sculptured tower to a very simple process—allowing it to proceed at the rate of one 10,000-sq-ft floor a week. Because of the unusual shape of the building—which provides panoramic views of the mountains—pie-shaped forms were required for the circular nodes. The three shear walls with their contiguous columns were poured using a single 45-ft-long combination form. A split cylindrical form was used for the three remaining individual columns. The use of precast stairs eliminated forming within the building.

All the forms were "flown" every Saturday by a crew of nine men who completed stripping and resetting in a little over three hours. The forms were aligned by means of a laser level to give an accuracy of within 1/8 in.

On Monday and Tuesday the ironworkers installed the reinforcing steel. On Wednesday the concrete was placed for the 9-in. floor slab and on the remaining two days, concrete was placed for the walls and columns. All erection was handled by a climbing crane at the center.

Among the other advantages of the construction method are: 1) quality of concrete work is extremely high—paint finishes only; 2) accuracy of alignment is advantageous for erection of the stick-and-panel metal skin; 3) "unlimited" horizontal shapes with repetition of floors are possible; 4) the cycle is limited only by the setting of rebar and inslab services.

Architecturally, the floor plan is compact with minimal area for servicing and circulation functions. The open structural plan along the sides of the triangle allows conversion of two typical two-bedroom apartments to a combination of one three-bedroom and one single-bedroom apartment configuration.

Patrick E. Loukes, who was chief architect of DMJM-Northwest during design of the building, and is now a vice president with the builder, William Simpson Construction, emphasizes that for the potential of systems such as this one to be fully exploited, collaboration in the project programming is a must between the design disciplines and the builder. For example, it is much easier to make adaptations to the structure to facilitate the system before the design has proceeded too far.

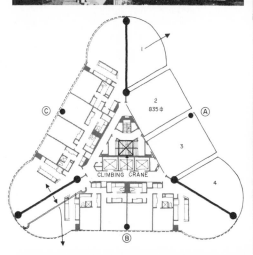

78

Steel system structure for high-rise housing

This project makes amply clear that systems building doesn't have to involve strikingly new concepts or "hot-out-of-the-lab" technologies. But rather it means closer engineering design; attention to practical details with respect to weathertightness; proving the effectiveness of different concepts and new uses of materials to building officials; and closer attention to scheduling, construction, and purchasing factors.

A new, modular steel-frame building system had its first large-scale U.S. application in 1972 in a $10-million, 458-unit, low- and moderate-income housing project, called Lake Grove Village. Located on an 8-acre urban renewal site on Chicago's South Side, it consists of three 10-story buildings and five 3-story buildings. The structure for the 10-story buildings was erected in 30 days, and enclosed with exterior panels, windows and roof in 18 days.

The structure is simple, straightforward and easily accommodates all utilities

There are only a very few elements—all standard—that comprise the structural system: 1) columns erected in two- and three-story lengths; 2) trussed-frames in U-shaped bays for wind bracing; and 3) bay-size floor panels made up of bar joists, open-web perimeter girders, and gypsum-panel floor decking. The floor panels were pre-assembled on two assembly lines at the site, trundled to the building location by straddle buggy, lifted to position by crane and simply bolted to column brackets. The floor elements include fascia panels; and also floor channels and head sections to receive the exterior wall panels that were installed simply from the inside. The floor deck served as a work platform, so no planking or internal scaffolding were required.

Interior columns are enclosed within party-wall partitions, and the column-width spaces between wall panels provide a natural location for mechanical chases for pipes and ducts. Sides of the wall panels facing the chase are covered with gypsum board for fire protection.

Lake Grove Village, Chicago, Illinois. Architects: Environment Seven, Ltd.

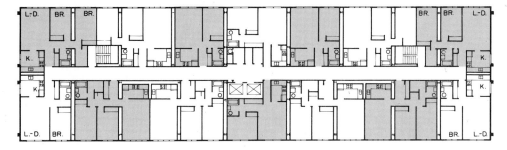

The structural elements are simple, standard, and thus endlessly repeatable.

The 10-story high-rise has four different bay widths. Where the interior columns occur they are enclosed by gypsum board, and the remaining space serves as a mechanical chase for pipes and ducts. The wind bracing consists of truss-braced columns and occurs at approximately the quarter points (note chases on plan).

There are no obstacles to interior layout, save for the few columns and chases

One obvious advantage of the system is the freedom permitted in space planning. The only fixed elements are the columns, vertical runs of piping and ductwork and stairs and elevators. Partitions could even be rearranged—within reason—to alter room sizes and shapes. Because the floor panels are framed with open-web members, there is plenty of room for horizontal runs of pipes, ducts and electrical conduit. Additional runs could be put in later, and it would be a simple matter to penetrate the floor system for the risers.

Perhaps one of the most salient features of the system is that there are no "specials" that have to be attended to and accounted for in working out a component list for a project. Because of the planning flexibility and the physical "openness" of the system there should be little temptation for architects and engineers to violate the discipline of the system—because the discipline, itself, from the mechanical standpoint is not rigid. And, of course, adhering to the discipline of a system oftentimes means the difference between whether there are significant cost savings or no cost savings.

A carefully detailed, highly disciplined kit of parts is the heart of the system

The system elements have to be fabricated and erected to closer tolerances than is customary with much of conventional construction. A study of tolerances as they affected system details was undertaken by the developers of the system, Component Building Systems, Ltd., in the construction of a prototype of the system—two duplex units—built on Chicago's North Side. Joint details of curtain wall panels were exhaustively studied and refined over a year's period. Component Building Systems took a direct hand in the Lake Grove Village project.

Component Building Systems, Ltd. functioned in many respects like the producers of industrialized building systems that have been imported here from Europe except in one very significant aspect—they did not need to set up a factory for manufacture of system components; the only such investment was for the on-site production lines for assembly of floor panels and these represented only very nominal cost. The steel frame elements can be bought conventionally; curtain-wall panels can be obtained from an established manufacturer; the gypsum planks are a standard product; the ceiling tile, however, are special cementitious units with a vermiculite-silicate base that are imported from France where the component building system now in use originated.

One of the changes made in the French system was to substitute conventional bar joists for the space-frame floor system which consisted of 7- by 11-ft panels made up of rods welded to rolled bars in a space frame configuration—like bar joists assembled in a saw-tooth pattern. This approach would have been uneconomical here, and, in fact the panels were costly to fabricate and transport even in France, so the system there has been changed to the U.S. arrangement.

The system originated in France in response to the need for more flexibility

The original system was first used, except for two small prototypes, for a 500-unit apartment project in Rouen, France consisting of 25 five-story buildings. It was developed by the Paris architectural firm of Lods, Depondt & Beauclair in collaboration with French building product manufacturers. Paul Depondt, who works as an architect both in France and the United States, and is currently a partner in the Chicago firm Architects International, says that the French government has constantly been increasing the minimum size of apartments, and because of the availability of larger spaces, some of the early post-war apartment buildings with fixed partitions have become less and less desirable—thus the incentive for more flexible partitioning.

Principals in Component Building Systems, Ltd. at the time this new system was developed were: Kenneth C. Naslund of The Engineers Collaborative of Chicago; Arthur O'Neil of the Chicago contracting firm, W. E. O'Neil Construction Company; and Arthur Bohnen, a Chicago building cost consultant.

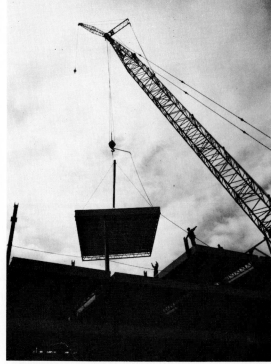

Floor panels are assembled on a production line that speeds fabrication and helps quality control.
The panels consist of bar joists, open web girders, gypsum-plank decking and metal fascia panels. Electrical conduit also is installed. A straddle buggy takes the finished panel from shed to building location. Panels are bolted to column brackets.

The wall system also consists of standard units—designed to be easily installed from the inside.

Wall panels are tilted up into place on top of a gasket which fits over an attachment that is part of the fascia assembly. The panel is held in place by a bolt at the head clip. The vertical joint between panels utilizes an interlocking detail and gaskets.

Both inner and outer skins of the panels are steel, with an acrylic finish on the outside and a vinyl laminate on the inside. The core of the panel is cementitious, mineral-particle material.

Exterior columns are fire-protected on the exterior with concrete that has been factory applied. Interior columns and trussed bracing have sprayed-on fire protection.

Floor covering is carpet applied directly over gypsum plank, or resilient tile over skin coat.

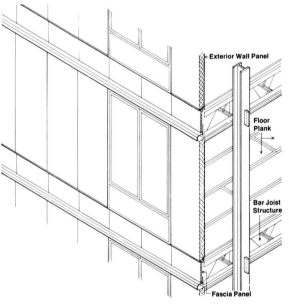

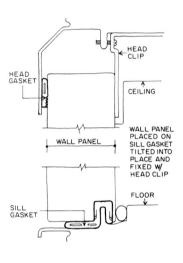

The mechanical and electrical services are easily installed in readily available space.

The open-web nature of the floor system allows plenty of room for conduit (left) and any necessary ductwork. The heating system for this project is perimeter hot-water convectors, so ductwork is minimum. Heating risers (center) and the plumbing wall (right) are concealed in a double-wall partition.

**Partitions and ceiling
could hardly be simpler.**
Within apartments partitions use a standard gypsum board system that leaves room for conduit (left). The ceiling title is a vermiculite-based product imported from France. A wedge between a clip on the tile and a clip on the bar joist holds the tile in place. A fire test for the floor assembly indicated 3¼ hr.

**The system won a turnkey competition
for New York state student housing**
Caudill, Rowlett, Scott together with Component Building Systems, Ltd. were winners in a $5.4 million competition to design and build student housing at the State University College at Brockport for the Dormitory Authority of the State of New York. Members of the team included W. E. O'Neil Construction Company, The Engineers Collaborative, and M. Paul Friedberg Associates, Landscape Architects.

The concept is a village of 1,000 students living in 200 two-bedroom, 25 three-bedroom and 25 one-bedroom apartments, with access to community centers, service facilities, parks, streets and plazas. The architects designed the project so that, "housing units serve as 40-ft deep 'partitions' modu-lated vertically and horizontally to create appropriately-scaled pedestrian ways."

The height of the buildings ranges from 1½ to 3½ stories above the street, and the apartment units are arranged horizontally in blocks of two to six units. Vertical circulation is via exterior stair towers.

CRS points out that the village-type layout incurs certain cost penalties that a typical double-loaded corridor, motel-type scheme does not. Exteriors of the apartment buildings are all weathering steel.

CRS notes that while the construciton process is quickened, and while the architect works on the project for a shorter period of time, there still remains a fixed amount of work the architect has to do that systems do not eliminate. There is a more concentrated effort in a shorter period of time—and there is vigorous involvement during the construction phase.

A significant advantage to the architect is that there is a quick turn-around process. Importantly, feedback is quick on how well the project works that can be plugged much sooner into future projects.

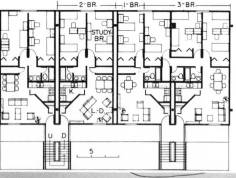

**Caudill, Rowlett, Scott won a
competition for this student
housing using the system.**
The $5.4 million project for 1,000 students illustrated here is under construction now at Brockport, New York for the Dormitory Authority of New York State. The massing of buildings and the site plan give a village-type character to the project. External stairs free the plan, cut construction costs.

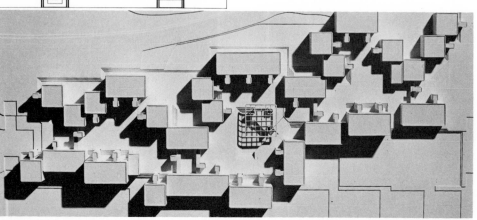

Prefabricated space trusses for recreational structures

Structural engineers Hirsch & Gray of San Francisco have designed a series of space-truss structures for several California architects that are straightforward and pragmatic, but at the same time evoke an esthetic response. Because the buildings have had a variety of shapes, the truss configurations have varied also.

The primary problem in the assembly of large-span, multi-member space structures has been to find some simple, inexpensive and repetitive way to connect many members into and through a typical joint. The approach developed by structural engineers Hirsch & Gray is to shop fabricate three-dimensional truss units, to ship them to the site, and finally to field bolt them together to form the completed structure. This approach permits the tailoring of individual members and connections to meet the actual forces imposed upon them, and also eases the geometric constraints. Field connections, made by bolting at the nodal intersections of units, can be easily varied as to number, size and type of bolt to suit individual load-transfer requirements. Detailing methods typically employ field connections across low-stress surfaces, with as much stress transfer as possible between units being taken in direct bearing between members.

Benefits of shop prefabrication of large units include minimization of field erection time, and less possibility of error in the field fit due to the greater tolerance control in the shop. Three-dimensional units—designed to be stable in themselves—provide inherent stability for the partially completed structure, requiring only minimal temporary shoring. Also, shop welding is less expensive and more dependable than field welding.

All the structures described here, were analyzed with the aid of computers, whose capabilities made practical the computation of stress conditions in these highly indeterminate structures under a wide variety of loading conditions.

The example shown on the first four pages is the roof structure for the base lodge for a ski area at Kirkwood Meadows at an elevation of 7800 ft above sea level in the California Sierra Nevada range. The architects, Bull Field Vollmann Stockwell, wanted an expression for this new ski complex that would have its own unique identity—enhancing the glamour and excitement of skiing. The severe snow load in this area required a roof structure that could support a design live load of 250 lb per sq ft —well over 3 million pounds for the entire roof. Furthermore, the site is some 100 miles from the nearest major city, and is covered with snow from late November to early July, leaving only four months' field time for construction.

Faced with these problems, the engineers sought a structure that could be fabricated mainly off site in a material and configuration that would satisfy both architectural and structural demands. the steel roof truss design shown here met these criteria.

The 8-ft-wide trusses are spaced 16 ft on center. This skipping minimizes material and the total number of connection points, and allows an interesting rhythm to be developed at the supports. The support "trees" are paired columns, 8 ft on center, linked in a prefabricate rigid frame unit for transverse stability—with a typical two-bay truss unit on top, turned at right angles to the main span trusses and supporting them.

Because of the pressures of time, design and fabrication took place almost simultaneously. The entire shop fabrication took four weeks from time of first mock-up and approval to shipment. Shipment and erection of the truss units on the previously erected columns took only two days—exclusive of final placement and tightening of all the field bolts. This was accomplished despite the fabricator's great concern that the required tolerances for bolt holes and general fit were too severe.

At a structural weight of 10 lb per sq ft, the load to weight ratio of the completed structure, designed to carry the high snow load cited earlier, is 25:1.

In designing space structures such as this one, the engineer must take care to simplify geometry, detailing and connections—repeating detail types as much as possible. Hirsch & Gray believe that if the truss system can be designed easily, it also can be built easily. They point out that attention should be paid to shipping dimensions and erection crane capacities as they might affect prefabricated module proportions and over-all size. The engineer also must think about how to minimize field shoring, and, *very importantly*, about the stability of the partially completed structure. Finally, Hirsch & Gray emphasize that the structural engineer must work closely with the architect and the owner to interpret architectural requirements and esthetic intent in practical structural forms.

Decision to use a space-truss roof system for Kirkwood Meadow lodge in the ski country of California's Sierra Nevada range was a natural outgrowth of design and construction considerations. These included: the doubtful availability of laminated timbers, a four-month construction season, a site 100 miles from the nearest city, a design snow load of 250 lb per sq ft, a very short design time, and the architect's intent to use the revealed structure as image. The prefabricated steel semi-space trusses took advantage of the long-established shop fabrication capabilities of the steel industry, and permitted shipment by road over two mountain passes.

The system employs 80-ft-long units (below) that span 40 ft from column to column, supported by inverted pyramids atop a column frame.

There are four space trusses (as the one above) comprising the roof structure for the lodge. All connections between the column frames, the support trusses, and the main-span trusses were made in the field using high-strength bolts. The 8-ft gap between top chords of the main trusses is spanned by 16-gauge corrugated metal decking, 3-in. deep.

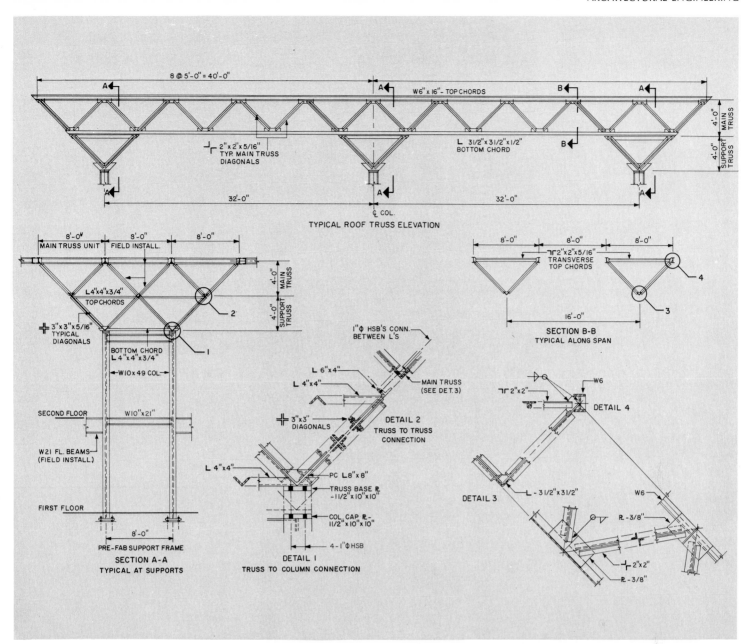

8 @ 5'-0" = 40'-0"

W6" x 16" - TOP CHORDS

⌐ 2" x 2" x 5/16"
TYP. MAIN TRUSS
DIAGONALS

L 31/2" x 31/2" x 1/2"
BOTTOM CHORD

4'-0" MAIN TRUSS

4'-0" SUPPORT TRUSS

32'-0"

32'-0"

⌐ COL.

TYPICAL ROOF TRUSS ELEVATION

8'-0"
MAIN TRUSS UNIT

8'-0"
FIELD INSTALL.

8'-0"

4'-0" MAIN TRUSS

4'-0" SUPPORT TRUSS

L4"x4"x3/4"
TOP CHORDS

2

⌐ 3" x 3" x 5/16"
TYPICAL
DIAGONALS

1

BOTTOM CHORD
L 4" x 4" x 3/4"

←W10x49 COL→

SECOND FLOOR

W10"x21"

W21 FL. BEAMS
(FIELD INSTALL.)

FIRST FLOOR

8'-0"

PRE-FAB SUPPORT FRAME

SECTION A-A
TYPICAL AT SUPPORTS

8'-0"
8'-0"
8'-0"

⌐⌐ 2"x2"x5/16"
TRANSVERSE
TOP CHORDS

4

3

16'-0"

SECTION B-B
TYPICAL ALONG SPAN

1"φ HSB'S CONN.
BETWEEN L'S

L 6"x4"

L 4"x4"

MAIN TRUSS
(SEE DET. 3)

⌐ 3"x3"
DIAGONALS

DETAIL 2
TRUSS TO TRUSS
CONNECTION

L 4"x4"

PC L 8"x8"

TRUSS BASE P.
-11/2" x 10" x 10"

COL. CAP P.-
11/2" x 10" x 10"

4 - 1"φ HSB

DETAIL 1
TRUSS TO COLUMN CONNECTION

W6

⌐ 2"x2"

DETAIL 4

DETAIL 3

L - 31/2" x 31/2"

W6

P. - 3/8"

⌐ 2"x2"

P. - 3/8"

Despite the intricate weblike appearance of the trusses, they have a right-angle geometry, permitting simple jigging for shop fit and assembly. Also, the planes of the bottom chord angle and that of the webs coincide, providing planar alignment for connections. The diagonal web members are doubled angles arranged in a star pattern for equal radius of gyration about both axes, and to emphasize the characteristic shapes of steel construction, the slenderness of the members, and the over-all pattern. In regions of high web shear, the angles are quadrupled. The top chords are wide-flange sections, sized for both local bending due to transverse loading between panel points, and direct compression from truss action.

Shipment and erection of the truss units on the previously erected columns took only two days, except for the final tightening of the field bolts. The trusses were stacked in pairs to form their own carriage, using a steerable trailer for support in shipment by truck from the fabricator's plant 100 miles away in Nevada.

Once at the site these units were set by crane atop "tree" supports and bolted, and were joined at the top chords by means of flanges visible in the photos below.

Prefabricated space trusses for two libraries

The library for York School in Monterey, California is basically one large mansard roof that touches the ground only along portions of three of the four edges. And to support the roof, rhomboidal-cross-section truss units were stacked on top of each other, at a 60-degree angle, as many as seven-tiers high. Similar truss units were used to close in the top. The units, composed of regular tetrahedra, were shop welded from small angles and tees.

Structural analysis, even by computer, posed a problem at the time because of the vast number of joints and members involved. The other problem facing engineers Hirsch & Gray was how to get an appropriate mating pattern between the truss units. They evolved a design that gave continuous and flat match surfaces, minimizing chance of field interference and misfit. Care in detailing and choosing member configurations, the engineers state, was the key to success in construction of the truss system.

The truss units were trucked some 120 miles to the site, and the entire erection took just seven days. Field connections between the truss units were made with two high-strength bolts at each of the panel points; a stitch bolt was used between panel points to connect webs of adjoining members and to prevent local buckling.

The building is 68-ft by 86-ft in plan, and 31-ft high. The front of the structure cantilevers some 27 ft beyond the side edge supports. In total, the space-truss units took about 40 tons of steel. They were 4 ft 6 in. on edge and ranged from 22 to 81 ft in length.

The library, designed by Smith/Barker/Hansen, architects, makes a feature of the exposed space-truss system. It was a recent winner of an Architectural Award of Excellence from the American Institute of Steel Construction.

Rhomboidal cross-section space trusses were stacked up like "steel masonry" to form the structural support for the huge roof of York School library. At the top of the building the units form a closed frame, but lower down there are cutouts to provide openings for entrances and windows. Field connections between truss units were made with high-strength bolts. The units were transported by truck 120 miles to the site where they were erected in seven-days' time. The engineers say that successful erection of the frame, with a minimum of field problems, could be attributed to careful study of details and member configurations.

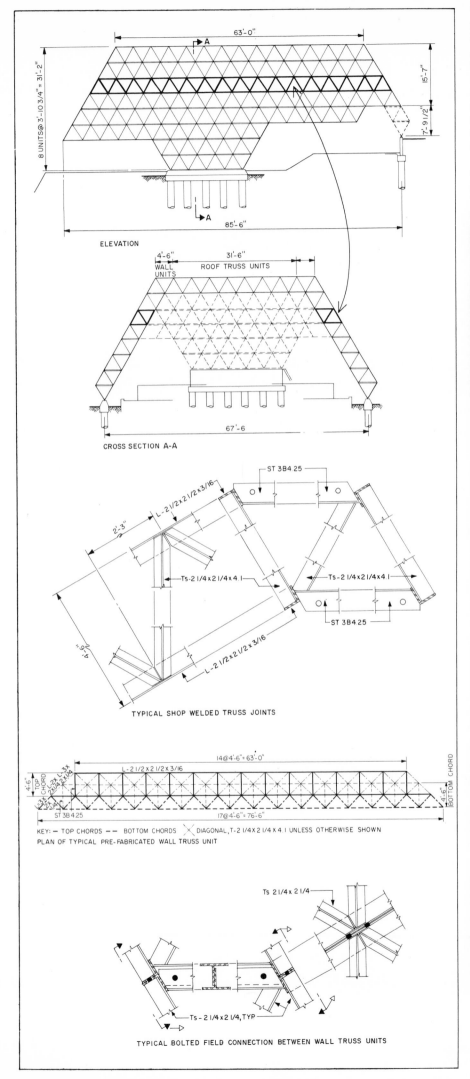

63'-0"

8 UNITS@3'-10 3/4" = 31'-2"

15'-7"

7'-9 1/2"

85'-6"

ELEVATION

4'-6"
WALL
UNITS

31'-6"

ROOF TRUSS UNITS

67'-6

CROSS SECTION A-A

ST 3B4.25

2'-3"

L-2 1/2 x 2 1/2 x 3/16

Ts-2 1/4 x 2 1/4 x 4.1

Ts-2 1/4 x 2 1/4 x 4.1

ST 3B4.25

4'-6"

L-2 1/2 x 2 1/2 x 3/16

TYPICAL SHOP WELDED TRUSS JOINTS

14@4'-6" = 63'-0"

L-2 1/2 x 2 1/2 x 3/16

BOTTOM CHORD

4'-6"
TOP
CHORD

L-3 x 2 L-3 x
2 1/4 x 2 1/4

L-3 x 2 L-2 1/4 x 2 1/4

4'-6"
BOTTOM CHORD

ST 3B4.25

17@4'-6" = 76'-6"

KEY: — TOP CHORDS — — BOTTOM CHORDS X DIAGONAL, T-2 1/4 x 2 1/4 x 4.1 UNLESS OTHERWISE SHOWN
PLAN OF TYPICAL PRE-FABRICATED WALL TRUSS UNIT

Ts 2 1/4 x 2 1/4

Ts - 2 1/4 x 2 1/4, TYP

TYPICAL BOLTED FIELD CONNECTION BETWEEN WALL TRUSS UNITS

Truss units vary in length from 22 to 81 ft to produce the irregular facades of the mansard-shaped building. About 40 tons of steel were used for the rhomboidal cross-section units. Field connections between units were made with two high-strength bolts at each of the panel points and with a stitch bolt midway between panel points to connect webs of adjoining members and to prevent local buckling. The space truss is painted and left exposed for visual interest.

Ephraim Hirsch photos except where noted

Kurt E. Oswald

The space-truss configuration for San Lorenzo, California Branch Library reflects a number of architectural requirements and conditions, among which were: 1) the architects' desire for a design statement for the main reading room; 2) a room shape that is an irregular hexagon; 3) a sloping roof line that allows a large clerestory on the building's east face. The one-story building was designed by architects Ostwald & Kelly.

The pattern is one of intersecting, elongated diamonds, the bottom-chord system being offset from the top-chord system by one-half module. The truss structure was painted yellow to make it stand out as an architectural element, and it serves as a "lighting fixture" as well, with fluorescent lamps installed in the longitudinal "V" formed by the bottom angle chord.

A number of practical factors also influenced the design by structural engineers Hirsch & Gray. They had to spend many hours developing details to provide the very narrow intersection angles. But they report that this effort was rewarded with rapid fabrication and erection—almost routine in its simplicity. The truss elements were shipped and erected within five days.

The structure was subdivided into 11-ft-wide by 80-ft-long units to meet the limitation of maximum-width unit that could be transported by highway. They were fabricated from angle and wide-flange sections. Despite the geometric complexity, only 48 field connections were required between the 10 truss units, themselves, and between units and Y-shaped columns. The first two modules were bolted together on the ground prior to lifting to form a stable unit; subsequent modules were erected singly, being stabilized by the one that went before.

Kurt E. Oswald

Geometry of the space-truss roof system reflects the plan shape itself—an irregular hexagon. The pattern is of elongated, connecting diamonds. The sloping units span from a wall at the back to Y-shaped columns in the front, the upper halves of the columns showing through the tall clerestory. Fluorescent lamps were installed within the bottom chords for indirect lighting.

The 80-ft-long truss units were fabricated from angle and wide-flange sections. Altogether only 48 field connections were required between the truss units and trusses and Y-columns. The 11-ft-wide truss units were bolted as shown in the drawing below. First step was bolting together of two units on the ground to form a stable assembly. Transportation and erection of the roof structure took only five days. One of the shop-welded connections is shown in the photo below. The intricate pattern of the final assembly is displayed in the photo at left.

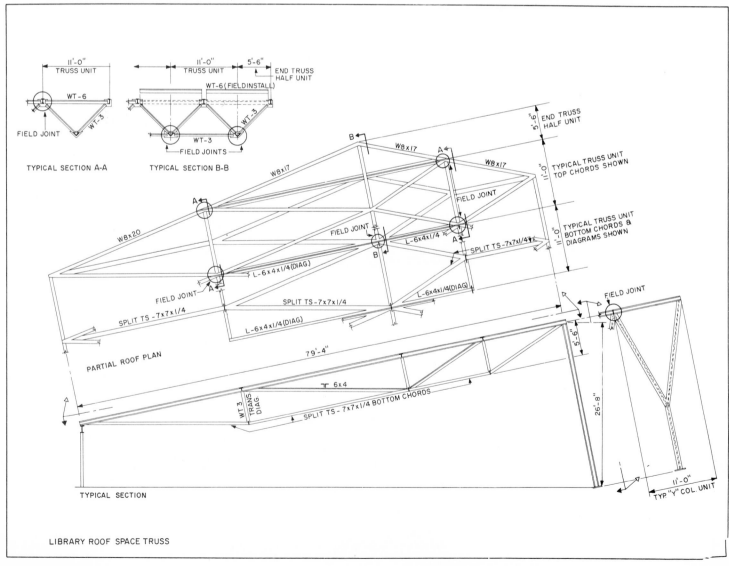

LIBRARY ROOF SPACE TRUSS

CHAPTER 4:

Structures—
Approaches to Suit Architectural
Design and Cost Requirements

This chapter presents the engineer as a problem solver with approaches ranging from the audacious to the pragmatic for projects that vary in scale, materials, aesthetic objectives and cost objectives.

Some of the solutions are unique, as is the 225-ft space truss with 5-ton steel connectors for the National Gallery's East Wing. Working on a smaller scale, engineers have helped architects achieve lofty, inspired spaces through inventive, neatly-detailed structures. In some cases the architect who has an innate sense for structure has suggested an approach the engineer was happy to develop. In other cases, innovative, but utilitarian, engineering approaches have produced considerable drama in large buildings and pleasant vistas in small ones. Engineered structures have doubled as architectural elements (for sunshading, as an example). In a more technological area, such as earthquake design, engineers have freed space for architects—and made possible construction in severe earthquake zones—by designing with steel plate walls in place of more conventional framing.

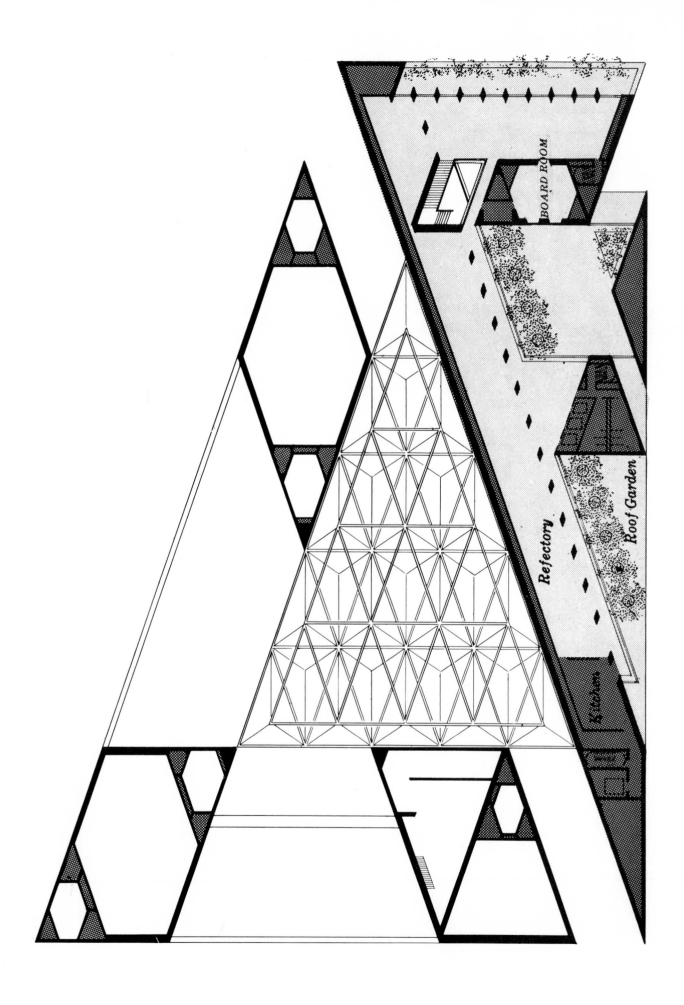

Space frame
for a national gallery

Superb execution is as important to a finished work of art as inspired design, a principle that is always reflected in the work emanating from the office of I.M. Pei & Partners. This principle is stretched very nearly to its limits at the National Gallery of Art, where extensive additions have been made. The building contains more than a few technical tours de force, among them a space frame measuring 150 by 225 ft with supporting members weighing as much as 5 tons, a 135-ft concrete girder only 4 ft deep, and cast-in-place concrete of extraordinary beauty.

The new building's triangular plan derives from its trapezoidal site at the intersection of Pennsylvania Avenue and the Mall. The main elements in the plan include a large office-library wing, three "pods" for exhibitions, and a central court, roofed by a skylighted space frame, for circulation and sculpture display.

The dimensions and angles of the plan's adjacent and overlapping isosceles triangles cannot, of course, deviate, and the fixed geometry made the integration—an architectural given—of appearance, structure and services a rigorous exercise, requiring constant cross-conferences between architects, structural engineers, mechanical engineers, contractors and suppliers.

Another factor, although it is not evident in either materials or craftsmanship, was budget. The $75 million in donated funds from Paul Mellon (this is a private gift to the nation) carried no carte blanche for the designers, and expenditures, which covered extensive underground facilities as well as the marble building, were carefully monitored for value.

NATIONAL GALLERY OF ART, Washington, D.C. Architects: *I.M. Pei & Partners—I.M. Pei; Leonard Jacobson, project architect.* Engineers: *Weiskopf & Pickworth* (structural); *Syska & Hennessy* (mechanical/electrical); *Mueser, Rutledge, Wentworth & Johnson* (foundation). General contractor: *Chas. H. Tompkins Co.* Construction consultant: *Carl A. Morse, Inc.* Space frame fabrication: *Chicago Heights Steel.*

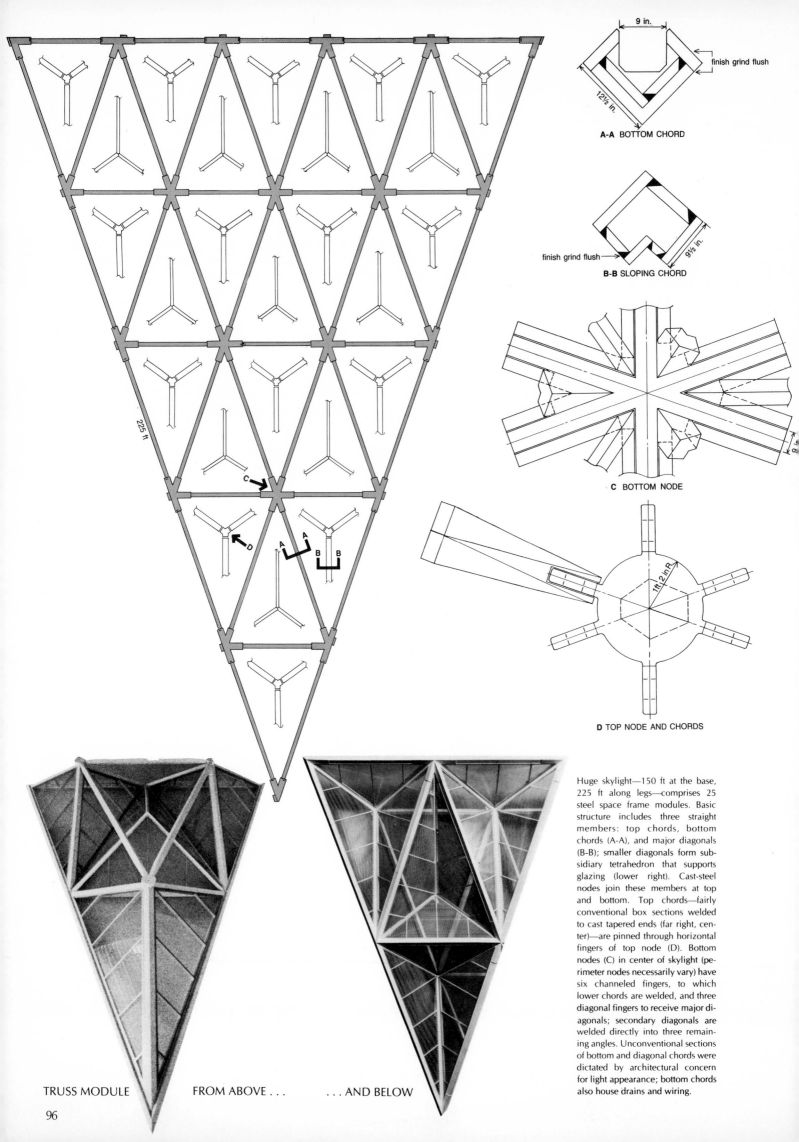

9 in.

finish grind flush

12½ in.

A-A BOTTOM CHORD

finish grind flush

9½ in.

B-B SLOPING CHORD

9 in.

C BOTTOM NODE

1 ft 2 in R

D TOP NODE AND CHORDS

225 ft

C

D

A A

B B

TRUSS MODULE   FROM ABOVE . . .   . . . AND BELOW

Huge skylight—150 ft at the base, 225 ft along legs—comprises 25 steel space frame modules. Basic structure includes three straight members: top chords, bottom chords (A-A), and major diagonals (B-B); smaller diagonals form subsidiary tetrahedron that supports glazing (lower right). Cast-steel nodes join these members at top and bottom. Top chords—fairly conventional box sections welded to cast tapered ends (far right, center)—are pinned through horizontal fingers of top node (D). Bottom nodes (C) in center of skylight (perimeter nodes necessarily vary) have six channeled fingers, to which lower chords are welded, and three diagonal fingers to receive major diagonals; secondary diagonals are welded directly into three remaining angles. Unconventional sections of bottom and diagonal chords were dictated by architectural concern for light appearance; bottom chords also house drains and wiring.

96

A-A

D

B-B

C

Space frame required three large steel castings: bottom node (C) (weight 5 tons, after machining), top node (D), and tapered ends of top chords (right). Straight members (A-A, B-B) are welded sections. Stainless steel pins fasten the top chords through fingers of top nodes. Skylight will rest on sliding bearings to accommodate differential thermal expansion of exposed and interior space frame members.

The beauty of the architectural concrete at the National Gallery originates partly from the remarkable formwork—one can justifiably call it cabinetwork—and partly from its material—the fine aggregate contains pinkish marble dust from the Tennessee quarry that supplied stone for both the new and old buildings.

On the main floor, ceilings above the entries and exhibition spaces are three-way coffered slabs. Along the Pennsylvania Avenue edge of the building, a concrete girder spans 135 ft along one side of the central court, and provides partial support for the large coffered slab at the main entrance (see formwork, with lighting track in place, top left, and coffers opposite, with girder at upper right corner). The depth of this long girder could not exceed 4 ft, the depth of the coffered slab, and is reinforced with exceptionally large bars.

The photograph directly above shows the inside of a coffer form, with conduit and electrical penetrations in place.

The architects wanted services in the large court to be invisible. From unobtrusive "air scoops" just beneath the skylight (lower left), high-volume nozzles supply air, which is returned—also invisibly—through "risers" in the grand staircase (below), through pits around several trees planted in the court, and through a planter.

As one of the building's structural engineers remarks, "People have to work very hard to make things look simple."

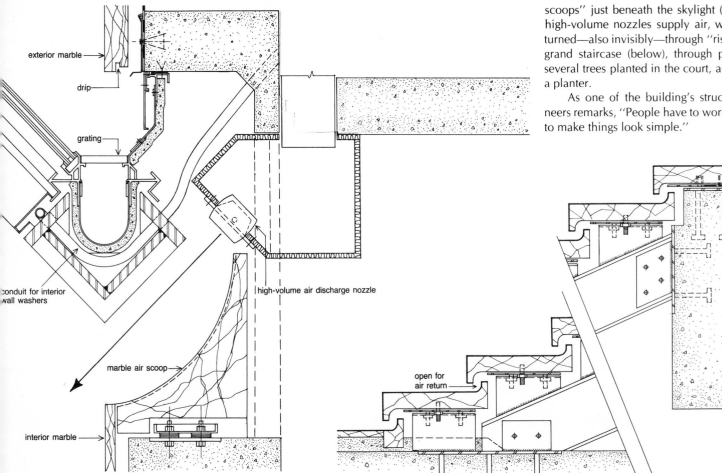

exterior marble

drip

grating

conduit for interior wall washers

high-volume air discharge nozzle

marble air scoop

interior marble

open for air return

SECTION THROUGH BOTTOM CHORD AT PERIMETER

AIR RETURN SUPPLY DETAIL

Long-span precast structure for a medical library

The problems a structural engineer expects to encounter with exceptionally long spans are compounded when those spans must carry exceptionally heavy loads. At the Medical College of Ohio at Toledo, the mix of problems included: library stacks supported on a 90-ft clear span; windows capped by 90-ft concrete beams; a combination of precast and cast-in-place concrete, with various deflections; and uneven loading above glass partitions.

The school's program for its new building called for three distinct facilities—a student-faculty union, administration offices, and a library with limited access. Moreover, the building was to be a symbolic entrance to the campus. The architectural solution places an office block at one end of the building and a separate union building tucked between a pair of stair towers at the other, with the library on the fifth floor bridging an open area between the other elements and serving as a lintel above the ceremonial gateway.

This solution places a heavy load—100 psf for library stacks, plus 50 psf for the floor slab—above an 80-by-90-ft void. The floor's 90-ft precast single tees are supported by two concrete Vierendeel trusses, 25 ft deep and 232 ft long, spanning the passageway. The Vierendeels are in turn supported by bearing walls at the stair tower and around the office block. Cantilevers project outside each truss to support carrels, me-

chanical space and exterior walls. Apertures in the Vierendeel web become doorways.

The great length of concrete elements in both directions created special concern for the effects of creep during curing and of thermal movement. The Vierendeels therefore bear on frictionless joints at the stair towers, allowing movement parallel to the plane of the trusses, though the bearings' stainless steel housings restrain transverse movement. Both ends of the 90-ft tees bear on Teflon-coated steel pads, one end carried on a sliding bearing, the other fixed but allowed to rotate.

In addition, the building has long expanses of glass both at the end wall windows and at partitions in offices beneath the library. Extremely careful analysis of loading ensured that deflection would not place intolerable stress on glazing (see diagrams below and on next page).

MEDICAL COLLEGE OF OHIO AT TOLEDO, Library, Administration and Student-Faculty Building, Toledo, Ohio. Architects: *Don M. Hisaka & Associates, Architects, Inc.* Engineers: *Gensert Peller Associates* (structural); *Evans & Associates, Inc.* (mechanical); *William B. Ferguson* (electrical). General contractor: *Rudolph/Libbe/Inc.*

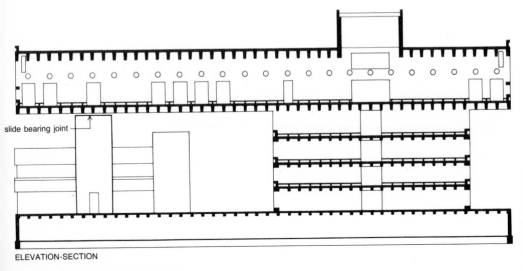

slide bearing joint

ELEVATION-SECTION

With an impressive show of structural audacity, a 90-ft concrete horizontal mullion surmounts a 90-ft butt-glazed window. Apparent vertical mullions for upper lights are in fact steel hangers. To protect glazing, close analysis of the end wall to calculate depth of initial camber and subsequent behavior of horizontal members was mandated by a

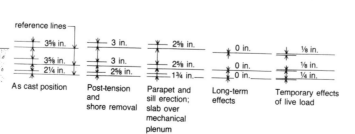

	As cast position	Post-tension and shore removal	Parapet and sill erection; slab over mechanical plenum	Long-term effects	Temporary effects of live load
reference lines	3⅝ in.	3 in.	2⅝ in.	0 in.	⅛ in.
	3⅝ in.	3 in.	2⅝ in.	0 in.	⅛ in.
	2¼ in.	2⅝ in.	1¾ in.	0 in.	¼ in.

number of considerations: deflection of the cast-in-place concrete must match as nearly as possible deflection of the precast tees at floor and roof; all three horizontal members required post-tensioning; deflection would increase as dead load was superimposed and as creep occurred.

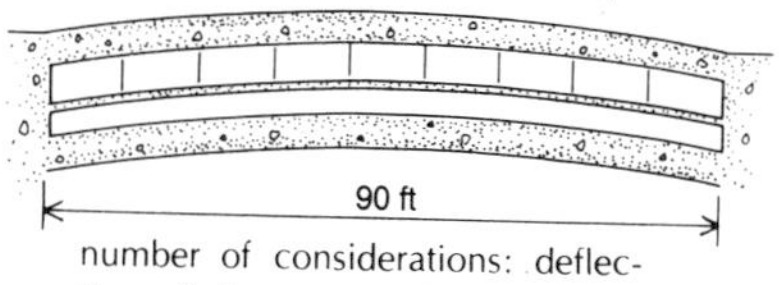

Robert E. Fischer photos

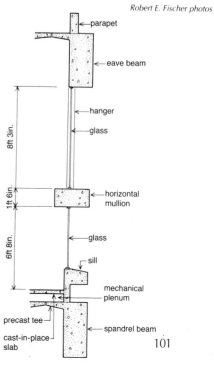

parapet
eave beam
hanger
glass
8ft 3in.
horizontal mullion
1ft 6in.
glass
6ft 8in.
sill
mechanical plenum
precast tee
cast-in-place slab
spandrel beam

Floors in office block are supported by concrete waffle slabs, topped by a plenum slab and left exposed on the underside. Eight-in. voids between waffles and plenum slab carry supply and return air in alternation, except at perimeter, where hvac piping is housed beneath removable flooring. Fluorescent strips for indirect lighting are inserted in coffers and between tees

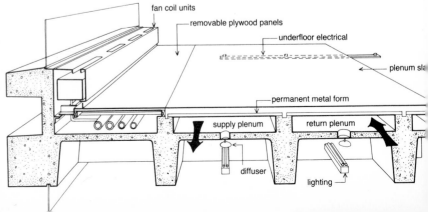

Inside face of Vierendeel truss is exposed at top of grand staircase to library floor. The girder conceals mechanical space behind, and is pierced by round outlets for air supply. Large exhaust grille beneath skylight leads directly to fan room supported by balcony tees. Passageway through the building (bottom right) is flanked by semi-detached student-faculty center and four-story office block. Glass-enclosed tunnel at rear runs to neighboring buildings. (For architectural coverage, see RECORD, August 1975, pages 86-89.)

While ordinary usage assumes overall uniform loading, here heavy loads are uneven, threatening windows and partitions in offices directly beneath library (above and lower right). Analyzing probable deflection of tees, engineers calculated separately magnitudes and locations of both sustained loads and temporary live loads, and designed reinforcement and post-tension accordingly.

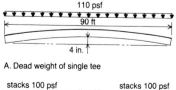

110 psf
90 ft
4 in.

A. Dead weight of single tee

50 psf
90 ft
3¼ in.

B. Superimposed dead load

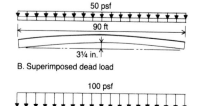

stacks 100 psf   reading 60   stacks 100 psf
90 ft
2⅛ in.

C. Partial stack loading

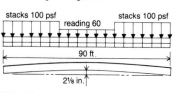

100 psf
90 ft
1¾ in.

D. Full stack loading

Structural engineering collaboration in architecture

The Denver-based consulting engineering firm of Ketchum Konkel Barrett Nickel Austin, with its New York affiliate Ketchum Barrett Nickel Austin/Besier, enjoys a widening reputation for sound, innovative solutions to tough structural problems.

That the reputation is earned is evident in the sampling of recent projects shown in the following pages. That it is relished is evident in the partners' shoptalk, which is laced with terms like "challenging," "exciting," and, pervasively, "fun."

That it is no accident is also evident. KKBNA's fruitful collaborations with architects spring first and last from a reservoir of exceptional engineering talent. But they flow too from organizational policies and patterns well calculated to nurture that talent.

These, say its beneficiaries, are a legacy of the firm's founder, Milo S. Ketchum, now retired but continuing to serve the firm as consultant, and his long-time partner, the late E. Vernon Konkel.

Michael H. Barrett, KKBNA's president, speaks of Ketchum as a teacher at heart, who saw the "practice" as just that: an ongoing effort to learn and hone skills. Believing that little learning derives from rote solutions to routine problems, Ketchum steered the firm instead toward the pioneering work in new methods, systems and products that is its stock in trade today.

Barrett is quick to point out that, for every structural tour de force, the firm engineers a dozen relatively unexceptional, and unsung, projects, adding that the firm's reputation for tackling the difficult and unusual can be a mixed blessing.

"It's discouraging," he says ruefully, "when a potential client hands a nice straightforward high-rise to another firm and tells us, 'There's nothing special about it. When I want a 500-foot free-form clear span, I'll come to you.'

"Sure we like the hard jobs, but even on ordinary jobs we have fun coming up with imaginative ways to save money, or speed construction, or do better justice to the architects's design."

Underlying KKBNA's openness to new concepts is a genuine empathy with the firm's architect-clients. The partners take pride in a cultivated ability to attune themselves to the architect's thinking and problems. And while convinced of the propriety of the architect's role as head of the design team, they do not see the role of the structural engineer as thereby reduced to "just a matter of designing connections and telling the client why he can't do what he wants to do."

Barrett likens the relationship between structure and architecture to a marriage, observing that while such elements as mechanical and electrical services can to an extent be divorced from —or appended to—the total design concept, structure and architecture are indissolubly wed.

... it is in the conceptual stages that the partners' time and talent yield the client the best value for cost. "A few clients have learned to get the most out of us, bringing us in early to explore all the possibilities, so the structure can really become a determinant part of the design. This, we love."

A similar analogy is drawn by long-standing KKBNA client William Muchow of Muchow Associates. "I use them for all my projects, right down to a creak in my living room floor," he says. "We work as if we were one office. . . My architecture is as much their structure as it is anything I put into it."

Such happy marriages don't just happen, however. Muchow emphasizes the value of continuity in his collaboration with KKBNA, believing that both parties gain by familiarity with the other's thought processes.

G. Cabell Childress, another self-described "old shoe" client, concurs, but inserts a practical note. At least in its work with frequent clients, he says, KKBNA encourages loose arrangements under which the firm undertakes the structural design on all of a given architect's project—large or small, simple or complex—for the same fee. "That way, if the architect really wants to go way out, KKBNA can afford to fly right along with him, and hope to make up any extra costs on future jobs."

Childress, himself not averse to the odd creative flight, likes to bring in KKBNA "as soon as I've determined 'what wants to happen' in the building, so we can work together to make sure what happens is contained in the simplest, most direct, most economical way."

Muchow agrees: "The best architecture comes from a team relationship where we stimulate each other to think creatively about basic objectives," but emphasizes that this is a process best begun in the beginning, before key design concepts are frozen.

To which KKBNA would add a hearty amen, pointing out that it is in the conceptual stages that the partners' time and talent yield the client the best value for cost.

"A few clients," Barrett says, "have learned to get the most out of us, bringing us in early to explore all the possibilities, so the structure can really become a determinant part of the design. This, we love."

Even so, KKBNA gets its fair share of problems posed in the mode of "the span is 327 feet and it has to be flat," as well as frequent commissions specifying structural forms—notably thin shells and space frames—that the firm has successfully executed in the past.

The partners accept the constraints of working within such pre-established design concepts philosophically. "Depending on how good the architect is—and a lot of them have a real feel for structure—things can often be done as well as if we'd been included at the beginning. An architect like Jim Ream, who has that innate sense and collaborates to the nth degree, winds up designing more of the structure than we do," Barrett says. "Our job isn't to originate design, but to refine the architect's concept and make it work."

At the same time, the partners are persuaded that too rigid parameters set too soon often add needlessly to the costs of a project. "We can figure out how to do almost anything if there's enough money," Barrett comments. "But we'd rather figure out how to do something better for less."

Fee schedules being what they are, KKBNA's appetite for adventure is not without impact on the firm's own finances. Barrett tells with a certain perverse pride of a recent "challenging" project entailing three stories and thirty-nine different floor levels. "On that one, our costs were two and a half times the fee," he relates, "plus twenty years of my life." But he adds, "If our main motive were to make money, we'd none of us be engineers to start with. There are all kinds of compensations."

The firm's official posture toward the profit motive is considerably less cavalier. While insisting on the superior satisfactions of creative work, KKBNA is also at considerable pains to maintain the efficient, productive—and profitable—organization needed to produce it.

Though formally a corporation, KKBNA operates as a partnership—and sometimes, says Barrett, "more like a commune." Of a staff of 105, 14 are partners, who share equally in decision-making, and many more are associates or project engineers, who also have a voice in the firm's management.

The resulting ratio of one chief for every three indians, the principals admit, is sometimes unwieldy. But it reflects their conviction that clients hire people rather than firms, and that the firm should therefore afford the people within it the fullest possible scope for action and initiative. KKBNA believes its relatively large number of relatively independent principals enables it to offer clients the best of two worlds: the personalized attention of a one-man shop, backed by the combined experience and resources of the firm as a whole.

Accordingly, every project is overseen by a principal from start to finish. On comparatively small jobs, the partner-in-charge may also act as project manager—and often project engineer and chief draftsman as well. More usually, though, the task of project manager is delegated to an associate or project engineer, with the remainder of the project team drawn as needed from a pool of engineers, junior engineers and draftsmen. In either case, the partner in charge retains full responsibility for over-all management.

Given the autonomy KKBNA partners enjoy, there is little rigid specialization within the firm. In principle, a deliberate effort is made to distribute work so that all firm members have opportunity to gain experience on a variety of project types. In practice, though, since jobs coming in do not always oblige by lending themselves to a random match with the people available to work on them, some of the firm's engineers have acquired more experience with certain types of structures than have others. And since some also have more interest in particular structures than do others, a certain amount of concentration has evolved among them.

Except in the case of the civil

engineering department, however—which now accounts for some 35 to 40 per cent of the firm's business—such specialization can be attributed more to happenstance than to planning.

The flexibility KKBNA espouses in its professional functions extends also to the management of the firm, which boasts no "business" partner as such. Rather, each principal is first a practicing engineer, and secondly a proprietor with administrative responsibility for certain aspects of the firm's operations.

President Michael Barrett, for example, also has major responsibility for long-range planning and business-development. Donavon Nickel, trained in business administration as well as engineering, manages personnel and finances, while David Austin handles general operations. And so through the roster of partners.

Since divisions of labor are based less on organizational logic than on the propensities and proficiencies of the several partners, KKBNA's smooth functioning is perhaps in some degree serendipitous. But is can also be credited to a shared philosophy that overrides the principals' individual differences in approach and emphasis. "We've all grown up together in the firm" Barrett explains. "We've learned to do things pretty much the same way, and we see things pretty much the same way."

Not surprisingly, in light of the import the principals place on

---

"When you get down to it, the only way to get business is to do a superlative job. We all seem to grow by taking on challenges. . . . We like to think we add to the profession by doing it—and we have our share of fun."

---

their common heritage, KKBNA's policies toward its younger members are geared to produce similar opportunities for fruitful apprenticeship and ongoing association.

"Ketchum," Barrett says, "looked at the shop almost as a prep school. He'd bring guys in fresh out of school, train them in the real world for a couple of years, and then suggest they go on for more experience. Well, we feel for people to operate at maximum efficiency for the firm, they almost have to grow up with us, so now when we've got good people trained, we do our best to keep them."

KKBNA, with its long tradition of learning while doing, regards the postgraduate education of young engineers as its proper province, but is sometimes less than happy about the extent of in-office training these new engineers sometimes require.

Because many teachers, however brilliant academically, have had little practical experience, "in some cases, they just don't know whereof they speak," Barrett observes. Moreover, engineering curricula are becoming increasingly theoretical. "The kids have so little time for applied courses, they never learn what all those theories are for. When they get out, they can't design a simple beam."

On the other hand, in hiring experienced people, "Either the experience isn't quite what you want, or you can't be sure it's what you want, or it's so limited you can't tell whether the man can do anything else. With recent graduates, their capabilities are a matter of record. Because they've learned to learn, it's easier to teach them our quality standards and ways of doing things. And it doesn't take all that long—in about a year we begin to get real productive work from them."

Thus while KKBNA occasionally takes on experienced engineers "out of necessity to get the work out," its preferred practice is to train–and retain.

KKBNA's emphasis on encouraging promising engineers to "grow up in the firm" is expressed in the variety of ways by which the principals contrive to give their younger colleagues ample growing room.

Among these, the team approach to project management is significant in that it enables engineers at all levels to participate meaningfully in a variety of projects. "We don't believe," says Barrett, "in sitting a man down to grind out the same calculation for three years. We like to move people around, give them all the experience they can handle, as fast as they can handle it."

Equally valuable is the obverse: the freedom allowed any staff member with a strong bent in a given direction to ride his particular hobby as far as it will take him. "If someone has a real desire to do something, we generally encourage him—and we don't ask for an economic feasibility study first."

Finally, the latitude for professional development that the firm tries to provide is backed by an open-door policy toward advancement within the firm. Barrett's assertion that "anyone who wants it and can do it can have my job" may be overstatement, but it is not lip service. KKBNA does in fact actively encourage its outstanding engineers to share ownership in the firm—as is borne out by the multiplicity of partners.

This open-endedness in organizational structure sustains KKBNA's goal of one-to-one relationships with clients, but the accompanying top-heaviness is not without drawbacks. "You can spread profits pretty thin," says Barrett, "and we sometimes have a problem finding new opportunities for people coming up."

As a result, KKBNA makes an active effort to grow steadily—without becoming muscle-bound in the process.

For example, while the firm's first branch office in Connecticut was opened largely because KKBNA (in the persons of Milo Ketchum and Rudi Besier) hoped thereby to work with some of the outstanding architects in that area, its metamorphosis as the present New York City office was seen as a way both to amplify that goal and to afford increased scope for the firm's upcoming professionals.

Similarly, KKBNA is now joining forces with other consultants to provide industrial engineering services on a team basis; anticipating the establishment—again with other specialist firms—of a design-build company in the mid-East; and exploring other possibilities for growth in capability.

The primary focus, however, remains on expanding KKBNA's traditional consulting practice, to which end the firm encourages all members in those pursuits—competitions, publications, teaching, professional and civic activities—which, with the standard promotional tools, add up to effective business development.

As such pursuits also contribute to the personal and professional growth of the firm's individual members, their fostering can perhaps be viewed as yet another example of KKBNA's penchant for enlightened self-interest.

"When you get down to it," Barrett sums up, "the only way to get business is to do a superlative job. We all seem to grow by taking on challenges, learning something new. We like to think we add to the profession by doing it—and we have our share of fun."

Fred Lyon photos

KKBNA AN AIRY SPACE TRUSS ROOFS A CALIFORNIA CHURCH AND AT ONE STROKE PROVIDES LIGHT, HEIGHT AND ORNAMENT

For the First Presbyterian Church in Berkeley, architect James Ream wanted a roof structure that would emphasize vertical height without overwhelming the intimate character he sought for the 670-seat sanctuary, and that would at the same time serve as a major decorative element for the room.

The 75-ft-square space truss that covers the sanctuary is described by the engineers as "a mesh of overlapping queen-post trusses with the bottom chords turned 45 deg to the top chords." Four steel wall trusses support the space truss around the perimeter; the wall trusses incline to extend the total roof span to 105 ft in both directions.

From the base of the vertical compression pipes, four tension rods, forming the bottom chord of the truss, run to the tops of adjacent compression members. These rods transfer gravity load from the center of the roof to the four steel wall trusses, and thence to the square concrete corner supports. (The oversized verti-

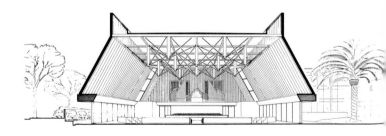

cal members house lighting fixtures).

The top chord of the truss is a grid of wide-flange sections, made continuous to resist vertical seismic force. The metal-deck roof diaphragm and diagonal corner beams transfer horizontal seismic forces to the corner supports via the wall trusses. Acoustical, rather than structural, needs determined the pyramidal shape of the ceiling panels.

FIRST PRESBYTERIAN CHURCH OF BERKELEY, California. Architects: *James Ream and Associates, Inc.,* Engineers: *KKBNA* (structural); *G. M. Simonson and T. R. Simonson Consulting Engineers* (mechanical/electrical). Contractor: *C. Overaa & Company.*

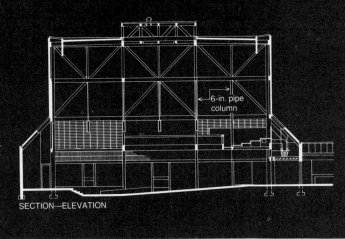

SECTION—ELEVATION

6-in. pipe column

RUGGED TRUSSES
DEFINE THE SANCTUARY
OF A CRUCIFORM CHURCH
IN THE ROCKY MOUNTAINS

The wood trusses at the First Presbyterian Church of Boulder, Colorado, represent something of an architectural about-face: the original scheme, which exceeded budget, was for a cast-in-place concrete structure with beams spanning 72 ft on each side. What eventually emerged from a series of conferences, models and design studies undertaken by architect William Muchow and structural engineers KKBNA is an exposed laminated wood-truss system that combines ruggedness with suavity.

The cruciform roof with its hipped corners is supported by two pairs of intersecting trusses, 12 ft deep and 72 ft long. The effective span, however, is only 48 ft; near the ends of each truss, a diagonal member extends past the bottom chord, somewhat in the manner of a knee brace, to reduce the length of the span and to act as a ridge beam for the hip roof. These beams span from the top chords of the trusses to a 13-ft-high masonry wall at the outside of the building, and supports the skylight over a corridor that circles the sanctuary. (Fixed louvers in sanctuary filter the sunlight.)

Interior surfaces are finished with 2-in. wood decking. In a mountainous area where heavy wind loads can be expected, this sheathing acts as a stiff diaphragm, transferring lateral load to the foundation.

FIRST PRESBYTERIAN CHURCH, Boulder, Colorado. Architects: *W.C. Muchow Associates.* Engineers: *KKBNA* (structural); *McFall & Konkel* (mechanical); *Swanson-Rink & Associates* (electrical). Contractor: *Wilkens Co., Inc.*

Robert E. Fischer photos

All truss connections have ½-in. plates on each side of the truss, fastened with 1½-in.-dia. bolts. Because different junctions required different connections—one joint receives 10 intersecting members—and because all connections are exposed, architects and engineers took special care in designing gusset plates for effective detail and pleasant proportions. Elements shown in the photograph above include one of the diagonal ridge beams and a vertical pipe, hung from the truss, that supports corner of organ loft.

mezzanine

organ
loft

glass panels

← skylight

skylight →

KKBNA IN DENVER, AN OLD-WEST SOD ROOF TOPS AN ENERGY-SAVING CONCRETE STRUCTURE

The mellow warmth of the rough-formed concrete walls in the offices of the Gary Operating Company stems, indirectly, from the owner's request for an energy-conserving building (the firm operates oil fields on contract). This request suggested thick concrete walls, which with their mass would retard heat gains and losses. Additional mass is provided by earth berms around much of the lower floor and by an old device of the frontier: a sod roof. (The grassy roof, clean of mechanical impedimenta, also offers an attractive view for a high-rise motel planned for a nearby site.)

An irregular plan resulted partly because architect Cabell Childress adopted an open plan, and partly because the building houses two related but autonomous firms (the second company appears here only in the exterior photographs). Although the plan dictated some irregularity of column spacing, as well as several curved and straight bearing walls, the structural system consists essentially of a 20-ft-square grid of columns supporting flat slabs.

In answer to an overburden of expansive clay on the site and to the heavy column loads—the roof comprises a 14-17-in. flat slab and 2 ft of earth—the structure is founded on straight-shaft caissons drilled down to clay bedrock.

OFFICES, GARY OPERATING COMPANY, near Denver. Architects: *Cabell Childress Associates.* Engineers: *KKBNA* (structural). Consultants: *Richard Weldon, AIA* (concrete). Contractors: *Al Cohen Construction Company* (general); *United Materials, Inc.* (roof construction).

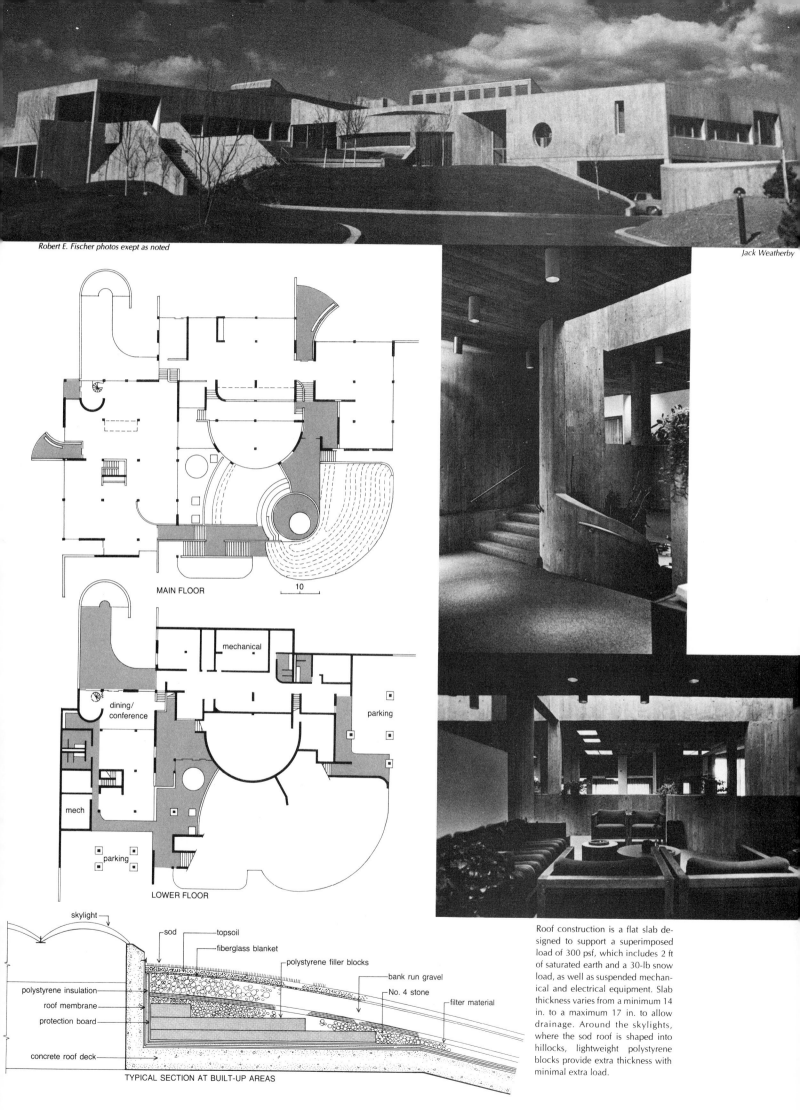

Robert E. Fischer photos exept as noted

Jack Weatherby

MAIN FLOOR

10

LOWER FLOOR

mechanical

dining/
conference

parking

mech

parking

skylight

sod

topsoil

fiberglass blanket

polystyrene filler blocks

bank run gravel

No. 4 stone

filter material

polystyrene insulation

roof membrane

protection board

concrete roof deck

TYPICAL SECTION AT BUILT-UP AREAS

Roof construction is a flat slab designed to support a superimposed load of 300 psf, which includes 2 ft of saturated earth and a 30-lb snow load, as well as suspended mechanical and electrical equipment. Slab thickness varies from a minimum 14 in. to a maximum 17 in. to allow drainage. Around the skylights, where the sod roof is shaped into hillocks, lightweight polystyrene blocks provide extra thickness with minimal extra load.

KKBNA ABOVE DENVER'S NEW ARENA,
A TWO-WAY CABLE TRUSS SYSTEM
SPANS AN AREA 300 BY 420 FEET
TO ROOF BASKETBALL AND HOCKEY FANS

The structural engineer's contribution to building design is nowhere more evident than in the large, unobstructed spaces needed for athletic events and their attendant spectators. For Denver's McNichols arena, which seats as many as 19,000 for basketball, hockey and concerts, KKBNA designed a two-way truss system to support a 300-by-420-ft roof. The trusses have 24-in. wide-flange top chords, 8-in. square tube verticals, and steel cable diagonals and bottom chords. Erection of the trusses went forward in three stages. First, the short-span trusses, including the diagonals, were positioned, and then the members of the long-span trusses were "woven" into place. Finally, the cables were tuned to bring the roof to its proper elevation. During erection, only short-span cables, which were over-tensioned, accepted load; the long-span cables were then tensioned to take load in the other direction and to relieve the extra tension in the shorter cables.

So that the building maintains a low profile on a prominent site, its lower portion is below grade, a condition that carried structural implications. Ordinarily, the basement slab in such cases acts as a horizontal strut to take lateral loads. Here, successive freezing and thawing of the hockey rink required that a continuous ¾-in. expansion joint circle the area. This in effect left a large hole in the center of the slab. The caissons that support the retaining wall and the seating bents were therefore designed to act as cantilever beams, taking horizontal forces as well as vertical.

The arena's exterior, sided with metal panels, takes its shape from the upper seating bents, which cantilever beyond the building line to form a broad rim around the roof. The hoods spaced around this rim house hvac ducts.

McNICHOLS SPORTS ARENA, Denver. Architects: *Charles Sink & Associates.* Engineers: *KKBNA* (structural). General contractor: *Centric Corporation.* Steel fabricator: *Zimmerman Architectural Metals.*

Robert E. Fischer photos below and above right

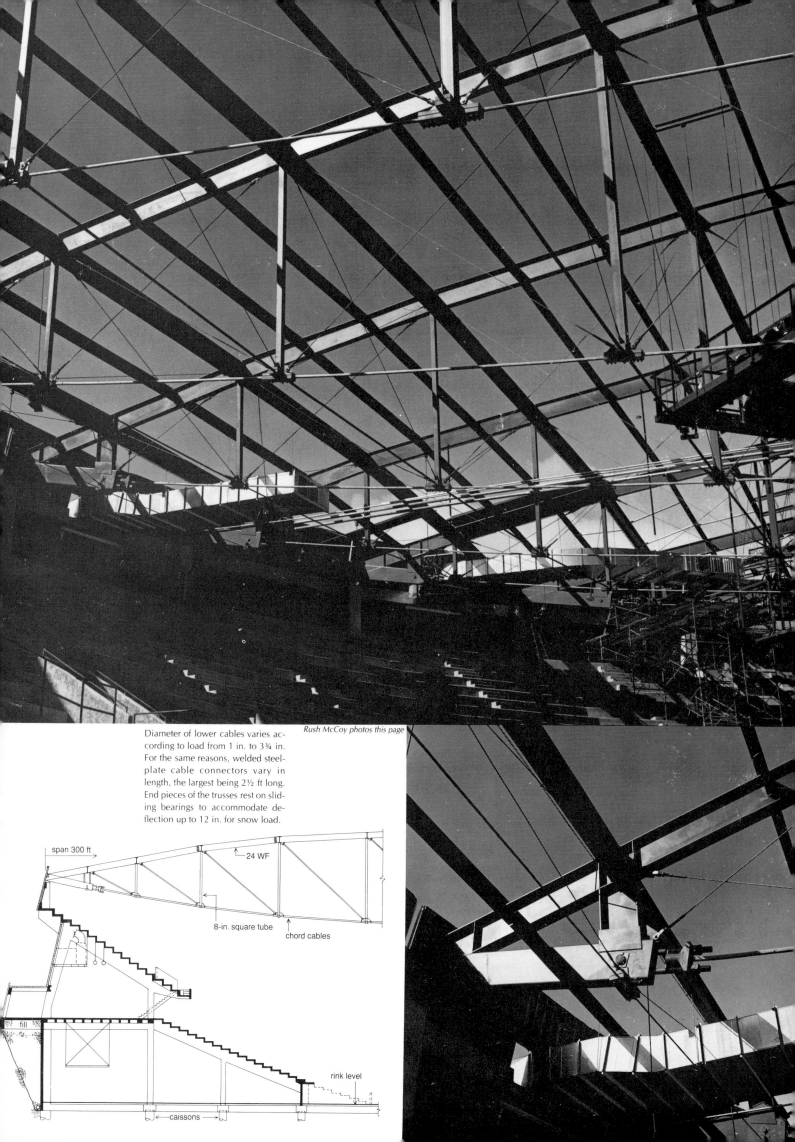

Rush McCoy photos this page

Diameter of lower cables varies according to load from 1 in. to 3¾ in. For the same reasons, welded steel-plate cable connectors vary in length, the largest being 2½ ft long. End pieces of the trusses rest on sliding bearings to accommodate deflection up to 12 in. for snow load.

span 300 ft

24 WF

8-in. square tube

chord cables

fill

rink level

caissons

KKBNA IN IDAHO, TRUSSED-ARCH ROOF SPANNING 400 FT ERECTED IN 26 DAYS

Erected in only five weeks from start to finish, the stadium roof at the University of Idaho uses a wood and steel trussed-arch system developed by KKBNA and the Trus Joist Corporation. The two-hinged arch spans 400 ft with a rise of 100 ft from the spring line to reach a maximum height of 160 ft above the playing surface; the building is 400 ft long.

The basic structural element consists of a steel-pipe web with top and bottom chords of *Micro-Lam,* very thin (1/8-in.) wood laminations. Six of the 2-ft-wide segments are bolted together to form an erection unit 12½ ft wide, 7½ ft deep, and 225 ft long from anchorage to the top of the arch. The segments are separated by 1-in spaces between clips, gaps that later serve as exhaust vents in the ceiling at the lower face, while the open-webbed arches serve as return-air plenums. Web members are typically 2-in. 16-gauge steel pipes set on 60-in. centers along segment seams, and staggered 30-in. for each segment within erection units.

UNIVERSITY OF IDAHO, Stadium Roof, Moscow, Idaho. Architects: *Cline, Smull, Hamill & Associates.* Engineers: *KKBNA* (structural). Contractors: *Trus Joist Corporation* (roof design-construct); *MacGregor-Triangle* (roof erection).

Dave Alfs photos

2 ft

2-in. dia. steel web

laminated-wood chord

A

A

angle clips

⅝-in. dia. bolts

SECTION A-A
bottom shown; top similar

Chords are spliced with 20-b...
24-in. bolted plates in a sta...
gered pattern along the length...
erection units (far left), wh...
steel-pipe web members al...
overlap at the splice (details se...
ond left). Pipes, flattened at to...
and bottom, are held by bolte...
angle clips on 5-ft centers alo...
each chord unit and staggere...
2½ ft from segment to segmen...
A 1-in. space is left open b...
tween clips on the lower su...
face; a urethane insulation ar...
elastomeric weather barri...
seals the top surface.

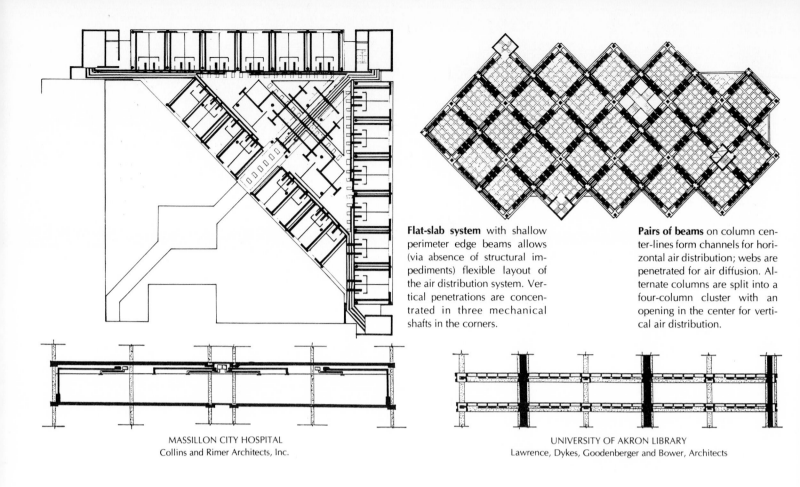

Flat-slab system with shallow perimeter edge beams allows (via absence of structural impediments) flexible layout of the air distribution system. Vertical penetrations are concentrated in three mechanical shafts in the corners.

Pairs of beams on column center-lines form channels for horizontal air distribution; webs are penetrated for air diffusion. Alternate columns are split into a four-column cluster with an opening in the center for vertical air distribution.

MASSILLON CITY HOSPITAL
Collins and Rimer Architects, Inc.

UNIVERSITY OF AKRON LIBRARY
Lawrence, Dykes, Goodenberger and Bower, Architects

Structural options for hospitals and schools

Like several idea-generating, resourceful structural engineers, Richard M. Gensert first thought of becoming an architect (as, curiously enough, did his partner, Miklos Peller). and when he did become an engineer, he developed the talent—as an architect —of recognizing and visualizing the total objectives of a building project in a way that lets him explore alternative ways of solving problems.

Gensert's office, Gensert Peller Mancini, located in Cleveland, excels in offering to its clients options (both design and cost) to choose from, as exemplified on these six pages. Many times, architects request that Gensert explore system design possibilities (as was the case with the Kent State School of Art on pages 118 and 119) or cost implications (as was the case with the study for Don Hisaka's hospital on page 120 and 121). Some other studies done by the firm include: 1) an evaluation of nine different steel-frame designs for a 25-story apartment building in Harrisburg—variations of interior columns placement were studied for gravity and lateral loads; 2) optimization of a framing system for the Hoover Industrial Plant in North Canton, Ohio, to allow crane runways and mezzanine areas to be interchanged; 3) a comparison of six different types of steel framing for high-rise office buildings.

Occasionally, on its own initiative, the Gensert office will develop structural schematics and preliminary cost studies for use in a) presentation meetings with architects and owners or b) early discussions between the architect and his engineering consultants. The offering of structural options to owners can be a powerful tool when an architect/engineer team is being interviewed by an owner. In Gensert's experience, clients respond favorably to an offering of several approaches, in preference to being locked in to one specific approach to structure that the engineer may have become identified with.

At job meetings with the architect and mechanical engineers, Gensert's people often present options for the routing of ductwork within the structural framing. Looking at this problem from the structural engineer's viewpoint, the firm may, for example, indicate an optimum configuration for duct layout that presents minimum interference and avoids costly fabrication for penetrations through structural members, or that takes consideration of the arrangement of structural elements to give sufficient depth for horizontal runs, or enough area for vertical shafts. More often than not, these suggestions coincide with good mechanical engineering practice. This approach, Gensert acknowledges, has been developed somewhat in self-defense. If

the structural engineer proceeds with structural design before the mechanical engineer has made a start, the structural engineer may end up doing a lot of redesign so as to accommodate the mechanical system.

Upon first encounter with this practice, the mechanical engineer may, understandably, show some reluctance for early involvement of this sort. He might, instead, prefer to get engaged in later stages of the job—initially offering general requirements for ceiling-depth and shaft-area requirements—feeling he cannot afford to get involved until the design has been firmed up. But once used to early design participation, the mechanical engineer, says Gensert, usually finds that time and cost benefits accrue to him in conceptualizing a job early.

In the Gensert office, all the principals and associates are adept at quickly developing structural options for architects' consideration, permitting a lot of interplay of ideas and approaches within the firm. This interplay allows careful cross-checking of one another's work that can help avoid possibly troublesom problems in the field. There also is a very good rapport between the designers and the production department. In fact, the head of the production and detailing department, Richard Fujita, is an excellent designer.

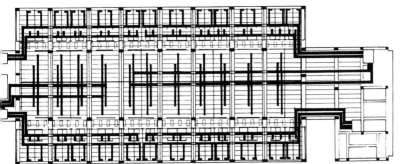

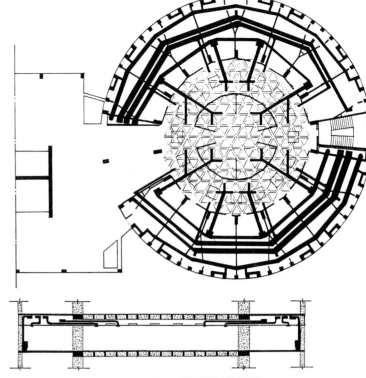

Concrete beams in the short direction have web penetrations for duct passage, located where stresses are at a low level. The location of openings was also governed by possible future changes in duct layouts, to preclude cutting openings later.

A structural depth of 24 in. for the 60-ft diameter clear-span core allowed only limited space for horizontal air distribution. The main supply and return ducts were concentrated within the band around the core where the structure was only 10-in. deep.

OHIO VALLEY GENERAL HOSPITAL
Highlands and Gilberti, Architects

CLEVELAND METROPOLITAN GENERAL HOSPITAL
Dalton vanDijk Johnson and Partners, Architects

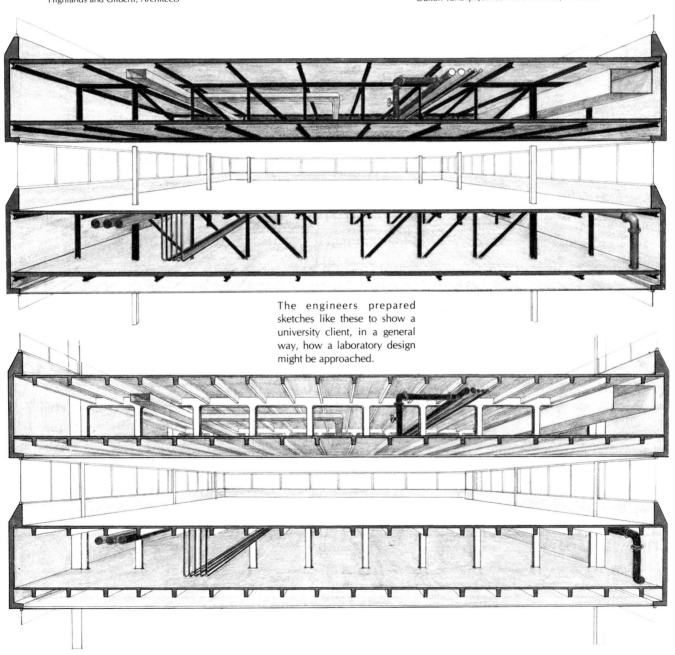

The engineers prepared sketches like these to show a university client, in a general way, how a laboratory design might be approached.

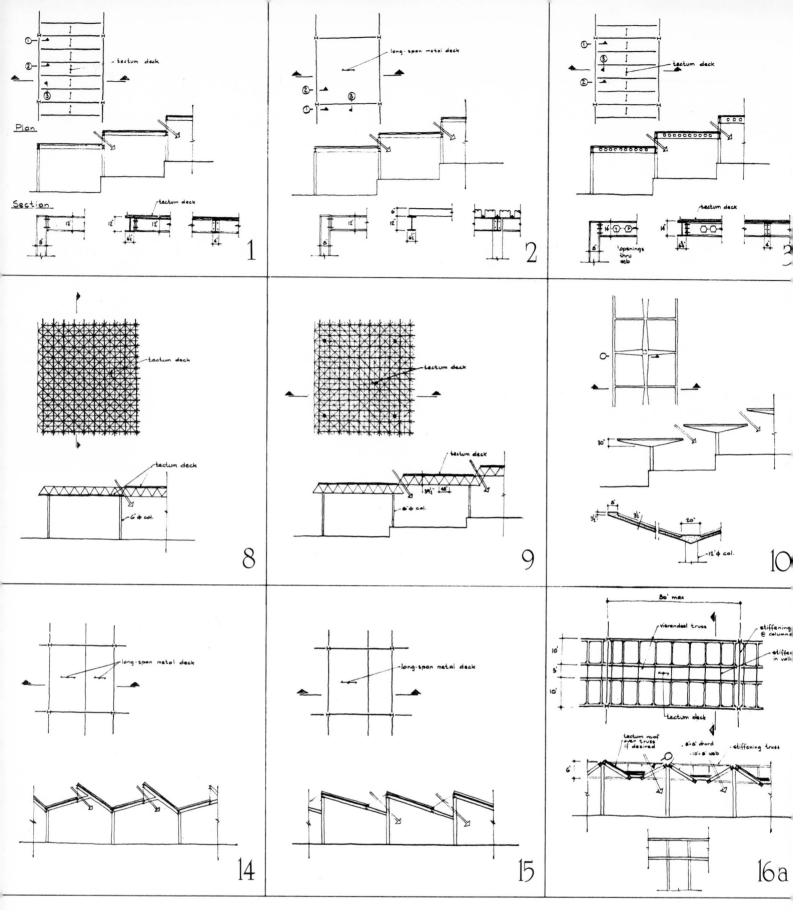

FRAMING	COST FACTOR	FRAMING	COST FACTOR
1	1.00	10	2.13
2	1.22	11	2.18
3	1.04	12	2.08
4	1.44	13	2.39
5	1.39	14	1.45
6	2.04	15	1.24
7	1.45	16a	2.55
8	1.47	16b	2.35
9	1.53	17	1.65
		18	1.55

The architect was given 18 schemes to consider

Even before architect John Andrews had a final plan for the School of Art at Kent State University (Ohio), he asked the Gensert firm to suggest some approaches to structure that would suit the site, a hill, and suit building use, which called for daylighting. The architects wanted a light, fun-type building with a structure that could be seen (so they had a concern about such aspects as intersection of materials).

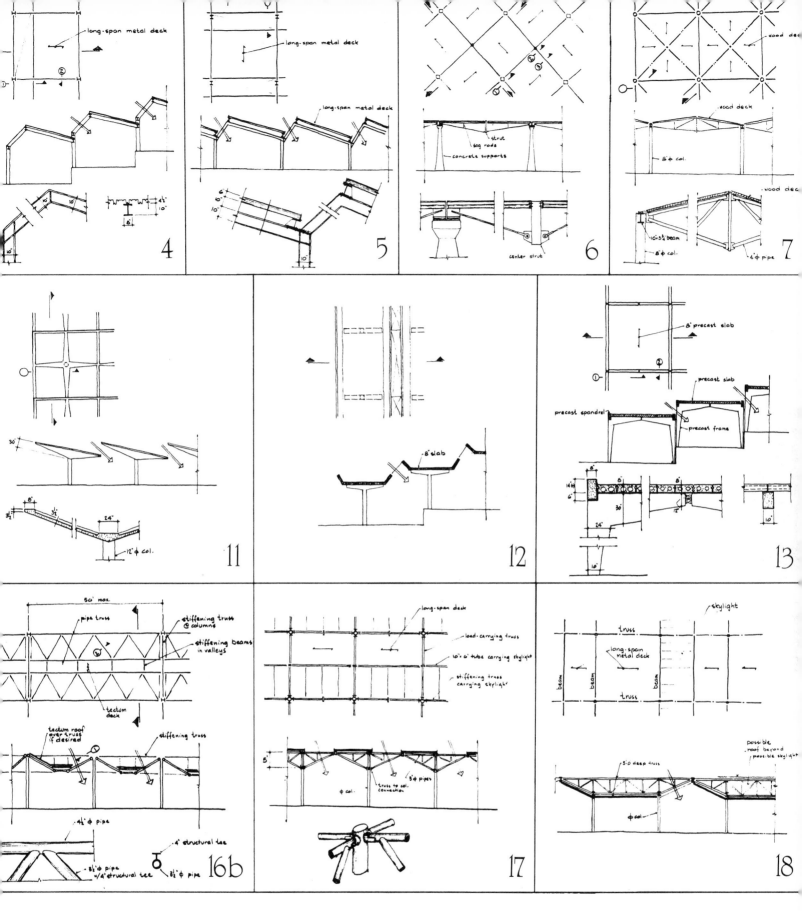

Though the building was on two levels, there always was exit to grade, so non-fireproof construction (such as exposed steel) coud be used. The physical relationship of site, building level and daylighting suggested a series of structural studies that set the stage for the final design. After the plan developed, the final configuration of the building actually took the shape shown in the photograph. The associated architects were John Andrews Associates and Ross Yamane Associates.

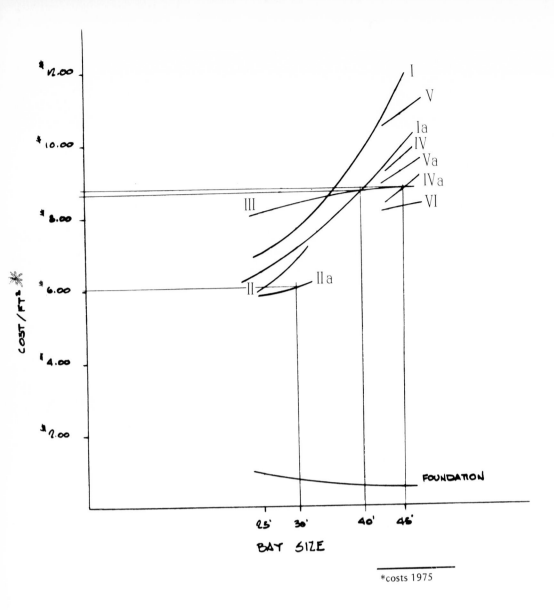

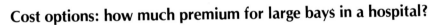

*costs 1975

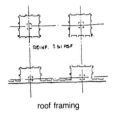

roof framing

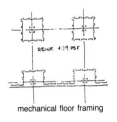

mechanical floor framing

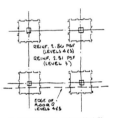

typical floor framing (3,4,5)

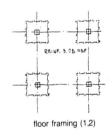

floor framing (1,2)

COMPLETE STUDY FOR
CASE I

Cost options: how much premium for large bays in a hospital?

That was the question architect Don Hisaka put to engineer Richard Gensert. The building in question was a new teaching hospital for the Medical College of Ohio at Toledo. Large spans were desirable to allow a considerable degree of freedom in the location of a modular system of wall and ceiling panels, and to permit arrangements of furniture and equipment.

The engineer determined that because of irregularities in plan and height of the building, a typical-bay study would not reflect true structural costs. For this reason he decided to study the structure on the basis of the building's full cross-section with different sizes of columns, spandrels, eave struts, roof and floors.

Column spacings of 20 by 20 ft, 30 by 30 ft, and 45 by 45 ft were investigated for structural steel and for the following concrete systems: flat slab, flat slab post-tensioned, coffered slab, coffered slab post-tensioned and precast concrete. Flat-slab systems, though economical, could not develop spans much beyond 30 ft, and thus had less flexibility from a planning standpoint. At the time of this study, for the spans studied, structural steel was more costly than the concrete schemes. Precast con-

crete showed a very uniform cost/span curve because of the many available sizes and types of components as spans varied. The systems ranged from cored planks to double tees.

A comparison of construction time showed precast concrete to take about one-half the time of cast-in-place concrete. But because the building was publicly funded, concern for financing costs was not a factor.

For bays up to 30 by 30 ft, a post-tensioned flat slab was most economical. For larger bays than this, the post-tensioned coffered slab became economical, but the cost increased rapidly, such that for bays above 40 by 40 ft, precast concrete became more economical. For the 25- by 25-ft bays the precast system was 8-in. cored planks; for the 30- by 30-ft bays the precast concrete was 8-ft wide double tees; for the 45- by 45-ft bays, three composite systems of precast and cast-in-place concrete were considered, using precast beam soffits that acted compositely with the cast-in-place slabs or with concrete fill. The latter, which was the most economical choice for the 45-ft bays, comprised precast double tee's, precast soffits, and poured-in-place fill.

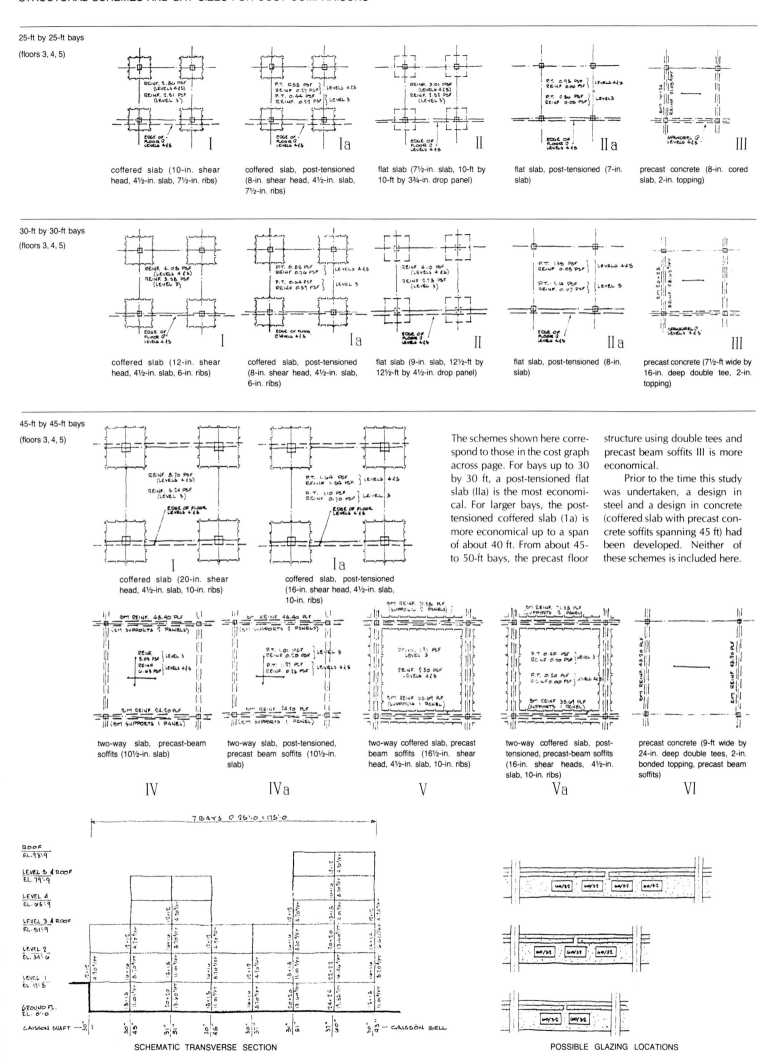

25-ft by 25-ft bays

(floors 3, 4, 5)

coffered slab (10-in. shear head, 4½-in. slab, 7½-in. ribs) — I

coffered slab, post-tensioned (8-in. shear head, 4½-in. slab, 7½-in. ribs) — Ia

flat slab (7½-in. slab, 10-ft by 10-ft by 3¾-in. drop panel) — II

flat slab, post-tensioned (7-in. slab) — IIa

precast concrete (8-in. cored slab, 2-in. topping) — III

30-ft by 30-ft bays

(floors 3, 4, 5)

coffered slab (12-in. shear head, 4½-in. slab, 6-in. ribs) — I

coffered slab, post-tensioned (8-in. shear head, 4½-in. slab, 6-in. ribs) — Ia

flat slab (9-in. slab, 12½-ft by 12½-ft by 4½-in. drop panel) — II

flat slab, post-tensioned (8-in. slab) — IIa

precast concrete (7½-ft wide by 16-in. deep double tee, 2-in. topping) — III

45-ft by 45-ft bays

(floors 3, 4, 5)

coffered slab (20-in. shear head, 4½-in. slab, 10-in. ribs) — I

coffered slab, post-tensioned (16-in. shear head, 4½-in. slab, 10-in. ribs) — Ia

The schemes shown here correspond to those in the cost graph across page. For bays up to 30 by 30 ft, a post-tensioned flat slab (IIa) is the most economical. For larger bays, the post-tensioned coffered slab (Ia) is more economical up to a span of about 40 ft. From about 45- to 50-ft bays, the precast floor structure using double tees and precast beam soffits III is more economical.

Prior to the time this study was undertaken, a design in steel and a design in concrete (coffered slab with precast concrete soffits spanning 45 ft) had been developed. Neither of these schemes is included here.

two-way slab, precast-beam soffits (10½-in. slab) — IV

two-way slab, post-tensioned, precast beam soffits (10½-in. slab) — IVa

two-way coffered slab, precast beam soffits (16½-in. shear head, 4½-in. slab, 10-in. ribs) — V

two-way coffered slab, post-tensioned, precast-beam soffits (16-in. shear heads, 4½-in. slab, 10-in. ribs) — Va

precast concrete (9-ft wide by 24-in. deep double tees, 2-in. bonded topping, precast beam soffits) — VI

SCHEMATIC TRANSVERSE SECTION

POSSIBLE GLAZING LOCATIONS

Earthquake design for a condominium

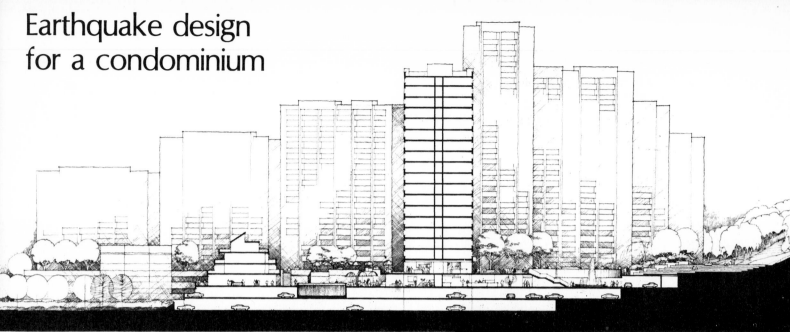

Early and continued collaborative-design input from the architect/engineer team led to economical structural and mechanical solutions for this 500-unit condominium in the East Bay area of San Francisco.

The team led by architect Harry C. Hallenbeck, sought a form that would provide visual cohesion from start to finish of the project. His studies resulted in the profile shown at the top of the page.

Early structural design was critical—minimum floor-to-floor height was necessary. The structural engineer determined that most economical was the use of reinforced masonry and conventional reinforced concrete shear and bearing walls combined with lightweight-concrete, post-tensioned, flat-slab floors. Seismic code limited buiding height to 160 ft. Layouts had to be studied to provide advantageously located shear walls.

Other important structural constraints included: 1) no drop beams could be used at mechanical chases or acrosss corridors; 2) corridor shear walls had to be separated from the floor slab during post-tensioning and "connected" later (to avoid induced stresses from slab shortening); 3) because the post tensioned slabs were not designed to carry the exterior masonry infill walls, these had to be "hung" between exterior bearing columns, or, in some cases, to be cantilevered from one of the bearing columns (see detail).

The mechanical and the electrical system include life-safety design. The project has a code-required automatic sprinkler system throughout. Elaborate fire-alarm, voice-communication and security systems have been provided, including a computerized surveillance system.

GATEVIEW ALBANY HILL, Albany, California. Architects: *Hallenbeck, Chamorro & Lin.* Engineers: *Shapiro, Okino, Hom & Associates* (structural); *G. L. Gendler & Associates* (mechanical). Consultants: *Cooper, Clark & Associates* (soils); *Bolt, Beranek & Newman, Inc.* (acoustics); *Sasaki Walker Associates* (landscape architects).

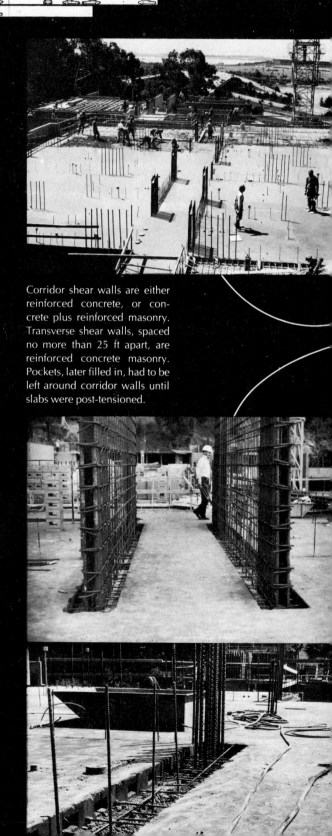

Corridor shear walls are either reinforced concrete, or concrete plus reinforced masonry. Transverse shear walls, spaced no more than 25 ft apart, are reinforced concrete masonry. Pockets, later filled in, had to be left around corridor walls until slabs were post-tensioned.

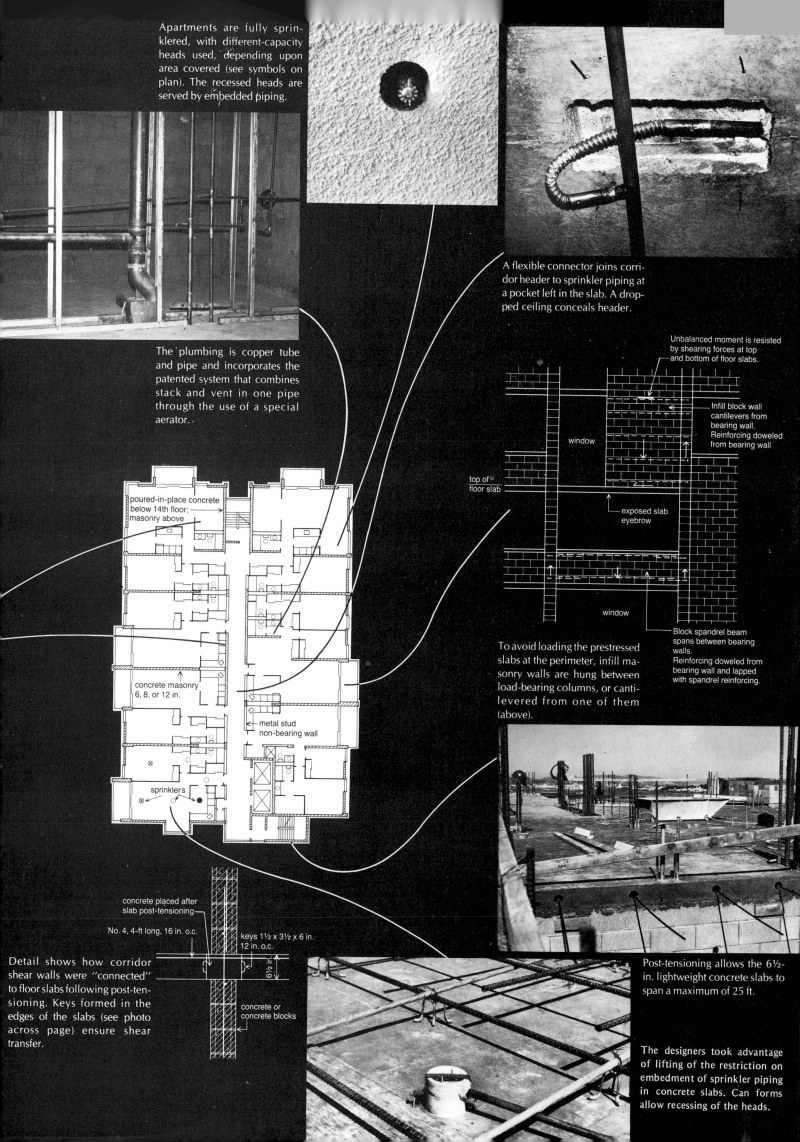

Apartments are fully sprinklered, with different-capacity heads used, depending upon area covered (see symbols on plan). The recessed heads are served by embedded piping.

The plumbing is copper tube and pipe and incorporates the patented system that combines stack and vent in one pipe through the use of a special aerator.

A flexible connector joins corridor header to sprinkler piping at a pocket left in the slab. A dropped ceiling conceals header.

Unbalanced moment is resisted by shearing forces at top and bottom of floor slabs.

Infill block wall cantilevers from bearing wall. Reinforcing doweled from bearing wall

window

top of floor slab

exposed slab eyebrow

window

Block spandrel beam spans between bearing walls. Reinforcing doweled from bearing wall and lapped with spandrel reinforcing.

poured-in-place concrete below 14th floor; masonry above

concrete masonry 6, 8, or 12 in.

metal stud non-bearing wall

sprinklers

To avoid loading the prestressed slabs at the perimeter, infill masonry walls are hung between load-bearing columns, or cantilevered from one of them (above).

concrete placed after slab post-tensioning

No. 4, 4-ft long, 16 in. o.c.

keys 1½ x 3½ x 6 in. 12 in. o.c.

6½ in.

concrete or concrete blocks

Detail shows how corridor shear walls were "connected" to floor slabs following post-tensioning. Keys formed in the edges of the slabs (see photo across page) ensure shear transfer.

Post-tensioning allows the 6½-in. lightweight concrete slabs to span a maximum of 25 ft.

The designers took advantage of lifting of the restriction on embedment of sprinkler piping in concrete slabs. Can forms allow recessing of the heads.

The materials represented in the following 17 pages run the gamut of steel, concrete, wood, plastics. The architectural forms derive logically from the nature of materials, but more importantly, the materials often are used multifunctionally. For example, the exposed steel frame on the following three pages has as its primary function resistance to wind and earthquake, but it also provides sunshades. The precast office building on page 131 deftly integrates the services in notched beams. The plastic spandrels in the Amex building on page 134 are not only enclosure—they support the glazing system as well. Two buildings in Texas and California have steel-plate walls to take the brunt of wind and earthquake. Factory-made utilitarian trusses were given a beauty treatment so they looked good exposed (page 132). The magic qualities of glass fiber for reinforcing plastics and concrete open up many new design possibilities (page 136). And an architect and concrete consultant get the most out of concrete by showing the builder how to form it (page 138).

Bare steel, a strong motif, resists earthquake/wind and carries overhangs to stop sun

Exposed seismic trusses—the main design motif for this building—were inspired, at least in part, by the ubiquitous gantry cranes, bridges, and other marine structures in Los Angeles harbor. The search for a nonrestrictive structural approach was stimulated by the need for an optimum area per floor of 30,000 sq ft and for a large expanse for open-office planning. The weathering steel braced frames at column lines, which are water-filled for fire protection, eliminate the need for shear walls either in the interior or in the window plane. This approach avoids the need for rigid frame construction, allowing the interior framing to have simple connections. Furthermore, the framing method permits an offset core. This, together with a continuous 7½-ft window height around the perimeter of the 10-ft ceiling, helps create a spacious feeling.

The large amount of glass used would not have been possible under the new California energy laws if steps had not been taken to shield it from the sun. The seismic frames provided means for attaching horizontal platforms to shade the glass and to permit window cleaning. This overhang is reduced to a 3-ft width on the north side so that a maximum of daylight can be utilized. On the east and west exposures, however, the full-width overhang is supplemented by a motorized shade that provides additional cutoff. Additional ambient lighting for the open-office floors is by pendant indirect HID luminaires.

LOS ANGELES HARBOR DEPARTMENT ADMINISTRATIVE OFFICE FACILITY. Owner: *City of Los Angeles, Port of Los Angeles Board of Harbor Commissioners—Fred Crawford, general manager; Joseph Taylor, architect.* Architects: *John Carl Warnecke & Associates—John Carl Warnecke, principal-in-charge; Emilio Arechaederra, project director; Edward Koester, project architect; Howard Kurushima, project manager; Edward K. Connors, Jr., project designer.* Engineers: *John A. Martin & Associates (structural); James A. Knowles & Associates (mechanical); Michael J. Garris & Associates, (electrical).*

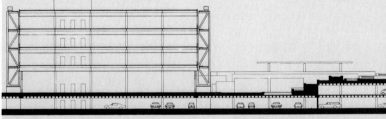

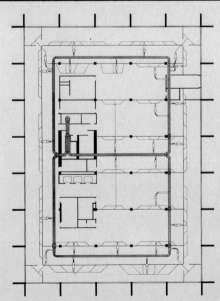

Horizontal forces caused by wind or earthquake are resisted by braced frames of weathering steel at column lines on 30-ft centers. The exposed, 60-ft long trusses, as long as could easily be transported, were welded to framing encased in concrete piers that provide plinths of varying heights, depending upon the grade.

All horizontal loads are redistributed at the second level, and shear walls transmit horizontal loads to the foundation (or reverse, for earthquake). While the office section is a steel structure, the parking garage is waffle slab.

Air distribution is single-duct vav to air-bar diffusers.

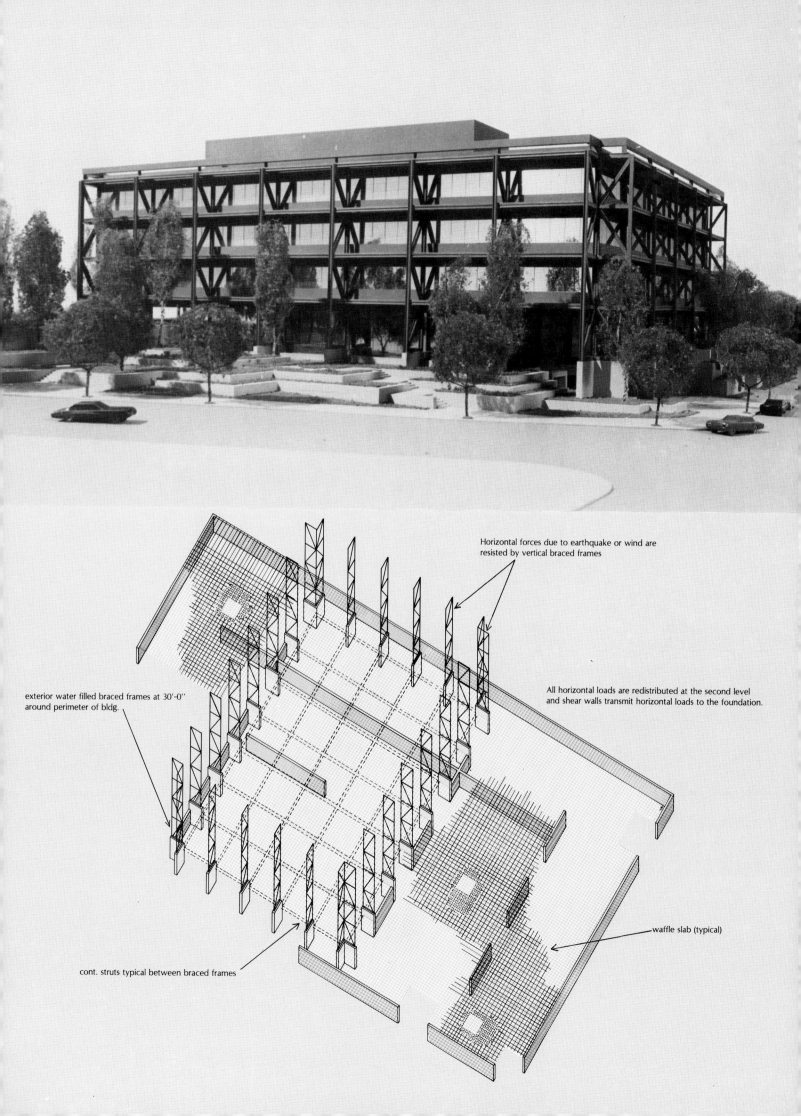

Horizontal forces due to earthquake or wind are
resisted by vertical braced frames

exterior water filled braced frames at 30'-0''
around perimeter of bldg.

All horizontal loads are redistributed at the second level
and shear walls transmit horizontal loads to the foundation.

waffle slab (typical)

cont. struts typical between braced frames

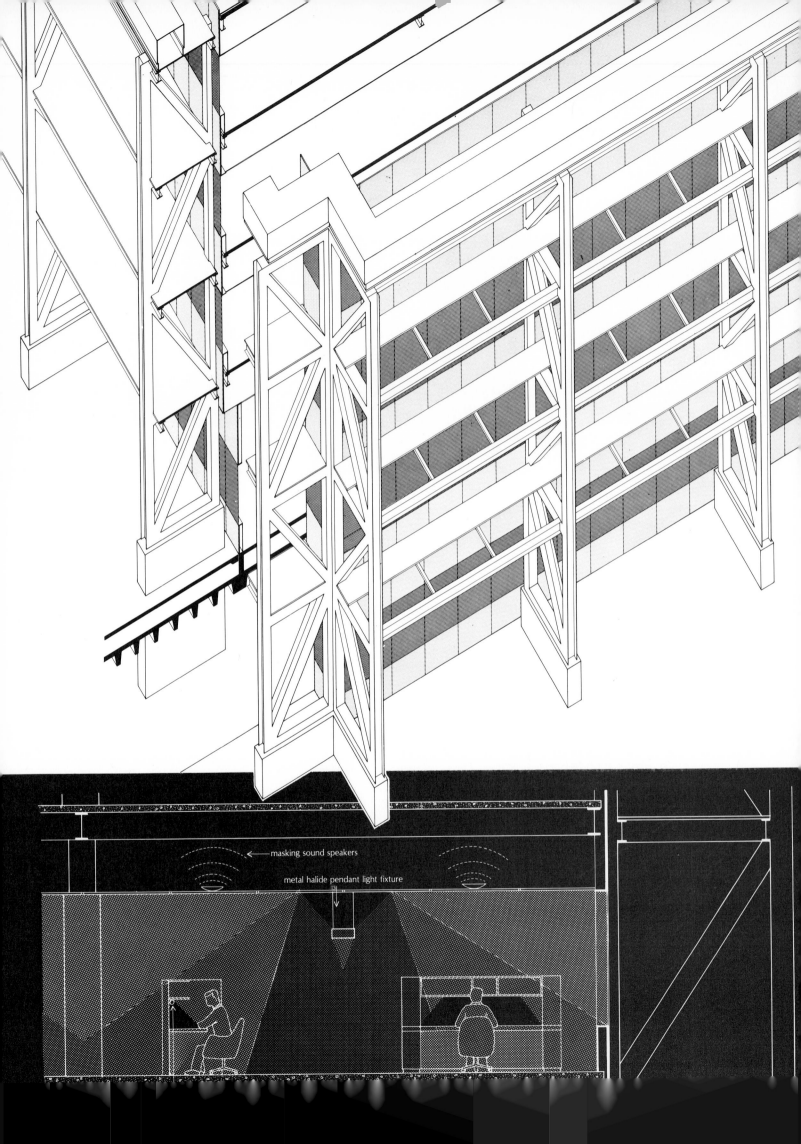

masking sound speakers

metal halide pendant light fixture

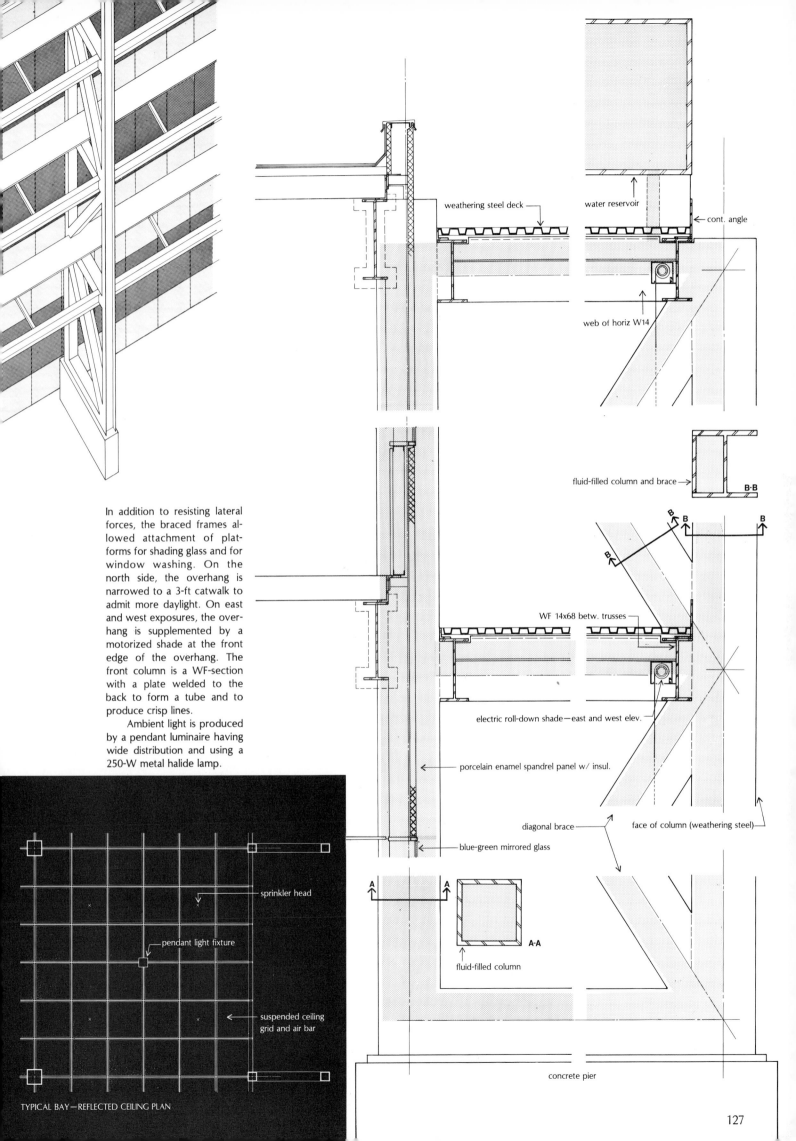

weathering steel deck

water reservoir

cont. angle

web of horiz W14

In addition to resisting lateral forces, the braced frames allowed attachment of platforms for shading glass and for window washing. On the north side, the overhang is narrowed to a 3-ft catwalk to admit more daylight. On east and west exposures, the overhang is supplemented by a motorized shade at the front edge of the overhang. The front column is a WF-section with a plate welded to the back to form a tube and to produce crisp lines.

Ambient light is produced by a pendant luminaire having wide distribution and using a 250-W metal halide lamp.

fluid-filled column and brace

B-B

WF 14x68 betw. trusses

electric roll-down shade—east and west elev.

porcelain enamel spandrel panel w/ insul.

diagonal brace

face of column (weathering steel)

blue-green mirrored glass

sprinkler head

pendant light fixture

A-A

fluid-filled column

suspended ceiling grid and air bar

concrete pier

TYPICAL BAY—REFLECTED CEILING PLAN

127

Steel-plate shear walls for a high-rise hotel

HYATT REGENCY HOTEL, Dallas, Texas. Owner: *Hunt Investment Corp.* Developer: *Woodbine Development Corp.* Architects: *Welton Becket Associates.* Engineers: Welton Becket Associates (structural); Herman Blum Consulting Engineers, Inc. (mechanical/electrical). General contractor: *Henry C. Beck Company.*

The Hyatt Regency hotel in Dallas, opened in 1978, is distinguished not only for the quality of its architectural design but also for the innovative and economical way the tower structure is stiffened to withstand the wind. The shear walls are steel plates—an approach that has long intrigued structural designers—rather than steel trusses or concrete walls.

The steel plate walls are located in the narrow east-west direction, where a conventional cross-braced frame would have encroached on the interior space. Conventional diagonal bracing was used in the longitudinal (north-south) direction for it could be

encased in the walls of the corridors. Diagonal bracing also was used on the lower two levels to allow openings for architectural design reasons. Wind shear is transferred from the shear walls to the diagonal bracing, spread through the floor diaphragm, and collected in the concrete foundation walls (see diagram top opposite page).

The steel plates are 10-ft high, 25.5-ft wide and 1-in. thick, and are stacked on top of one another between columns. In some cases two panels are connected horizontally to form a broader shear wall, and depending on the stress in the wall, two or three stiffeners were added to the panels. According to

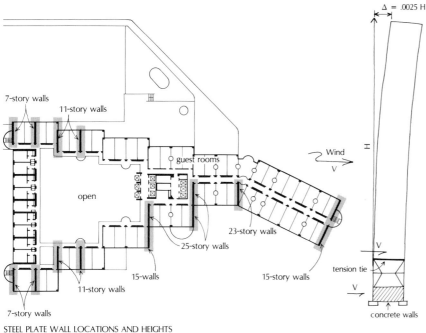

7-story walls
11-story walls
guest rooms
open
Wind
V
23-story walls
25-story walls
15-walls
11-story walls
15-story walls
7-story walls

$\Delta = .0025\,H$
H
V
tension tie
V
concrete walls

STEEL PLATE WALL LOCATIONS AND HEIGHTS

The steel plate shear walls (untreated, top, and fire-protected, center) use stiffeners according to the stress the wall takes. Outriggers (top) are for support of reflective glass curtain wall. Diagonal bracing (above) parallels the corridors.

Richard Troy, director of structural engineering for the architect, a 270-ft-high section in the tower will deflect only 8 in. at the top under the design wind loads.

The engineers chose steel shear walls rather than concrete because the contractor felt this would lengthen the construction time, and because a steel moment-resistant frame would have required a tremendous amount of steel and large members.

Because the design accommodates both lateral wind bracing and vertical forces, steel was saved by slimming down the columns and beams. The architects estimate a savings of approximately $2.85 million (1978).

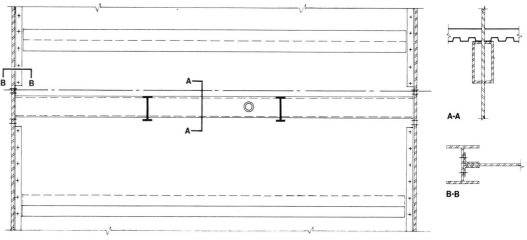

B B
A
A
A-A
B-B

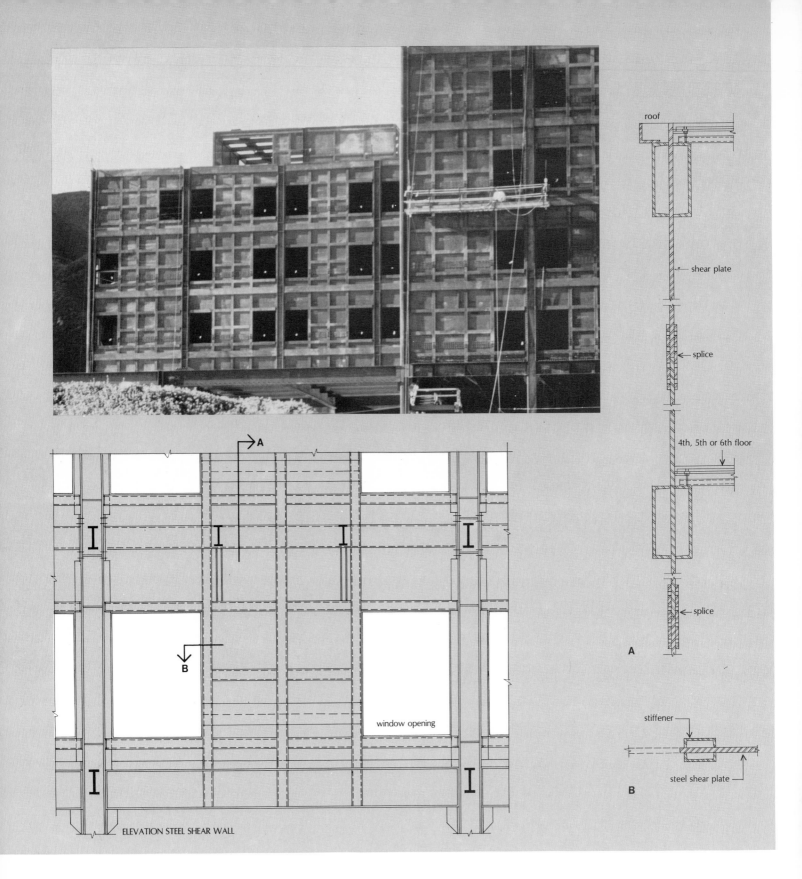

roof

shear plate

splice

4th, 5th or 6th floor

splice

A

stiffener

steel shear plate

B

> A

B

window opening

ELEVATION STEEL SHEAR WALL

Steel-plate shear walls for a hospital

OLIVE VIEW HOSPITAL, Sylmar, California. Architects: *The Luckman Partnership—Richard C. Niblack, project designer; Richard McKnew, project director.* Structural engineers: *Welton Becket Associates—Richard G. Troy, director of structural engineering; Robert Flick, project engineer.* Consultant: *Ralph Richard, University of Arizona, Tucson.* General contractor: *TGI Construction Corp.*

Built to replace the original Olive View hospital that was totally destroyed in the 1971 Los Angeles earthquake (and constructed on the same site for economic reasons), the new hospital has steel plate shear walls to meet strict earthquake codes and to permit more freedom in the interior spaces.

The four-story, cruciform-shaped building, set atop a two story rectangular base, was designed to meet a maximum credible earthquake ground acceleration of 0.69g. While a concrete building could have met this force, it would have required thick shear walls, usurping valuable interior space, and requiring an elaborate layout of reinforcing

steel around windows. On the other hand, a steel moment-resistant frame would have used 40 psf of steel.

The structural system consists of steel plate walls bolted to columns and girders, becoming an integral part of the structural frame (see detail) on the upper four stories, and concrete panels encasing the columns and beams on the lower two levels (so construction could keep pace with a fast-track schedule). Conforming to stress criteria computer-developed by a consultant for the steel plate walls, the walls at levels 3 to 5 were made ¾-in. thick, and at levels 5 to 6, ⅝-in. thick.

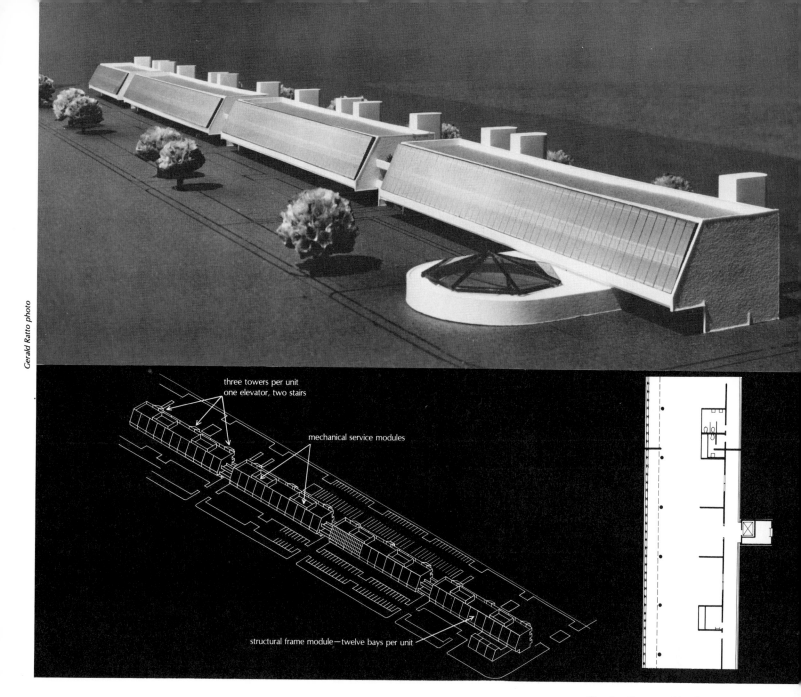

Gerald Ratto photo

three towers per unit
one elevator, two stairs

mechanical service modules

structural frame module—twelve bays per unit

Sloping glass wall for studio offices

OAKMEAD OFFICE CENTER, Sunny-vale, California. Developer: *Holvick, de Regt, Koering.* Architects: *Jacob Robbins and James Ream.* Engineers: *Ketchum Konkel Barrett Nickel Austin (structural); Climate Engineering (mechanical); Valley Electric (electrical); WTW Inc. (civil).* Contractors: *L. E. Wentz Co. (general).*

Lofty ceilings and a long sloping wall of glass create a spacious studio atmosphere in this speculative office complex 50 miles south of San Francisco. The north side of the Oakmead office complex, which is nearly a quarter-mile long, looks out over a lake in the Oakmead Village Industrial Park. Because of its orientation, architects were able to sheathe this entire side of the building in bronze glass, to allow a view and use of daylight without negatively affecting the air conditioning. The inclined glass extends office space out past the main ceiling, creating a skylight effect inside, and reflects clouds rather than wavy mirror images of nearby buildings outside. The complex is served by a single-loaded open corridor on the south side whose overhangs block solar gain.

To raise the ceilings and provide flexibility in air distribution, standard precast concrete beams were notched to provide room for ductwork within the beam. By using this method, and by raising the ceiling above the bottom of the precast beams for visual interest, ceiling heights of 9 ft 8 in. were possible. Cost was projected at $32 per square foot (1978).

Construction combined concrete that was site precast, factory produced, and cast-in-place. The south, east and west walls, which must resist high seismic loads, were panelized, allowing erection of one story at a time with a minimum amount of bracing.

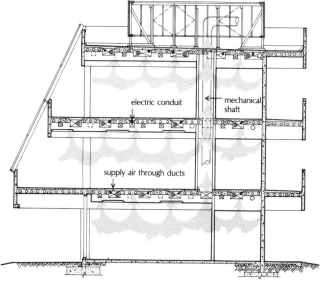

mech. penthouse

electric conduit

mechanical shaft

supply air through ducts

Lightweight wood trusses for a suburban office building

AMERICAN POLE STRUCTURES OFFICE BUILDING, Houston, Texas. Owner: *American Pole Structures, Combustion Engineering, Inc., Linn Tranquilli, liaison.* Architects: *Clovis Heimsath Associates— project team: Clovis Heimsath, Kerry Goelzer.* Interiors: *Clovis Heimsath Associates.* Engineers: *Nat Krahl & Associates (structural); Rex Bullock (mechanical).* Contractor: *RAYKO Construction Company.*

"A building with personality" was the instruction the client, a manufacturer of steel utility poles, gave architect Heimsath for their 8,000-sq-ft home office building in Houston. He responded with a building that neatly disposes three different types of space, giving prominence to the employees' lounge/ refreshment area, and privacy for executives in the upper portion of the split-level space.

Because he thought splayed wood trusses would suit the building program, Heimsath showed the client a golf clubhouse in which he had used prefabricated wood trusses for visual interest in roof and ceiling lines, and for creating skylighted space. They liked what they saw, so the architect developed the three ceiling configurations shown across page. All of the wood trusses were factory fabricated with *Gang-Nail* connectors fastening the webs and chords, most of which are only 2 by 4's. The trusses are supported by steel beams and octagonal shaped utility poles made by the owner. Though the owner wanted to display his product, the poles neatly serve as chases for wiring.

The building cost only $28 per sq ft (1978), and the architect attributes this to use of low-cost materials, exemplified by the trusses and by finish materials such as plywood and prefinished siding. Where the trusses are exposed over the lounge, care was taken in selecting lumber for appearance, and the bottom chords have 2 by 6's on either side to give a heftier look, to partially conceal the nail plates, and to provide a chase for wiring.

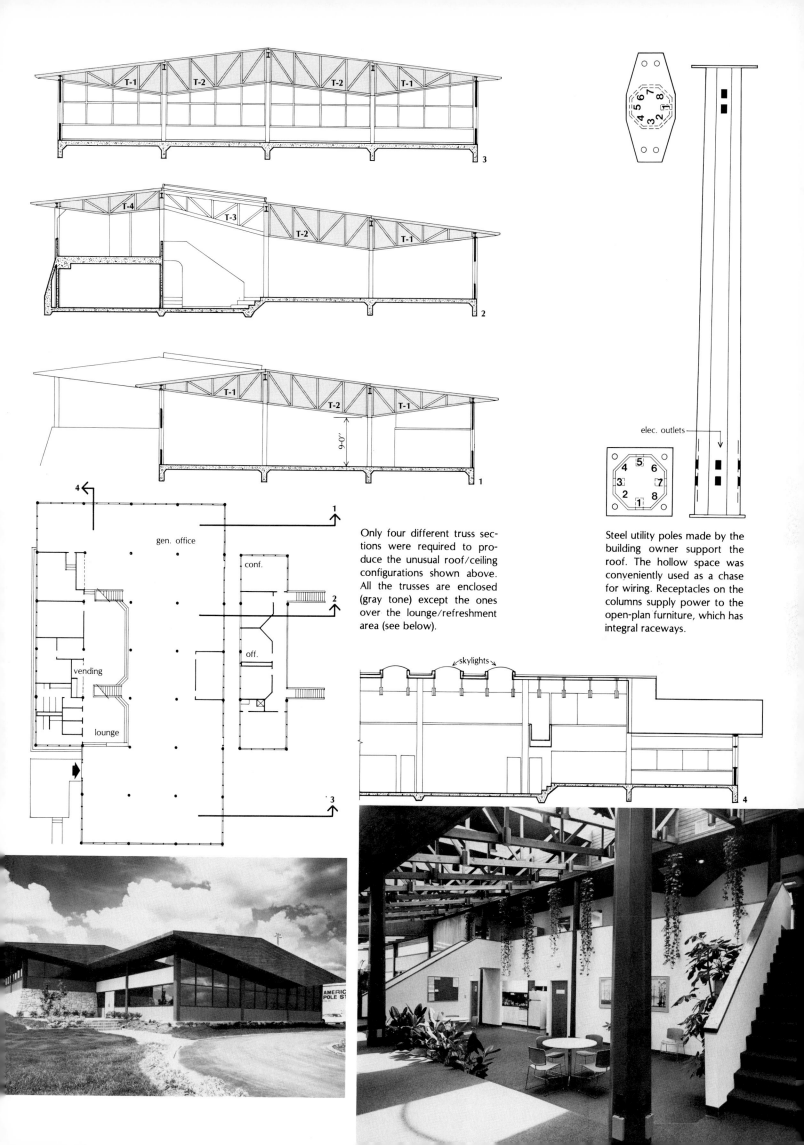

Only four different truss sections were required to produce the unusual roof/ceiling configurations shown above. All the trusses are enclosed (gray tone) except the ones over the lounge/refreshment area (see below).

Steel utility poles made by the building owner support the roof. The hollow space was conveniently used as a chase for wiring. Receptacles on the columns supply power to the open-plan furniture, which has integral raceways.

Load-bearing plastic spandrel panels

AMEX HOUSE, American Express Headquarters Europe. Owner: *American Express Company—Alun Price, project design and construction manager; Arthur Goodwin, project coordinator.* Architects: *GMW (Gollins Melvin Ward) partnership in association with Peter Wood Partners.* Engineers: *Zinn Burgess and Associates (structural); F.C. Foreman and Partners (mechanical/electrical).* Consultants: *Building Programs International; Peter Hodge and Associates (FRP panels); Cushman & Wakefield; Peter Corsell Associates (curtain wall).*

Sheathed in sparkling white ribbed panels of plastic and bands of marine-blue glass, American Express' new Euopean headquarters sits on a podium overlooking England's famed resort on the English Channel at Brighton, and in its palette and detailing reflects some of the ambience of this centuries-old resort. The broad plaza in front offers a large public space, and this, combined with the building's siting and landscaping, provides a transition to the surrounding clusters of row houses. Further, the form of the building allows maximum office space without blocking sun and daylight from the surrounding buildings. The tower was splayed at the corners to enhance daylighting.

Amex House, which opened in 1977, is the company's largest facility outside the United States, bringing together operations of their three main travel-related services, which were handled formerly by a number of offices in the county of Sussex, as well as in London. Reasons for selection of the Brighton site were the availability of skilled personnel, good telecommunications, and its proximity to London—only an hour's train ride.

The spandrel panels, most of which span about 24 ft between columns where they are supported by corbels, not only look structural, with their rib and flange sections, they are, carrying the 6½-ft-high glazing system of butted lights of blue glass—laminated to obtain the blue color the architect wanted. The spandrel materials were selected to withstand the exposure of the sea-coast air, and to maintain a pristine whiteness.

Richard Einzig, BrechtEinzig Limited photos

The spandrels are stiffened by the ribs and by box sections formed in the flanges. The elegance of the spandrel section is enhanced by the vertical board effect on the webs. The flanges, which work as beams, are connected by 4-in.-square preformed tubes on the rear side used to clamp the panels to the reinforced concrete corbels.

The material for the spandrels is glass-reinforced polyester (GRP, as it is known in England). The backs of the panels are separated from the occupied space by means of fire-resistive insulating panels.

A butt-glazing system with structural silicone sealant has been used, with sealant bonding to a neoprene spacer attached to the mullion. Details show glazing and the rain baffle at the spandrel joints.

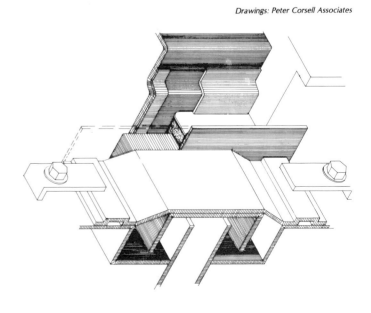

Drawings: Peter Corsell Associates

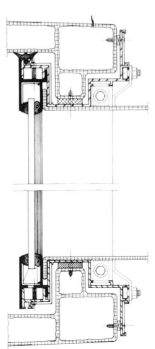

Fiberglass
plastic used for a house

For several decades, the technological glamor and the unique physical properties of plastics have intrigued architects, but not until the last few years have they been used in architectonic ways—taking on forms that mirror their structural characteristics and that are possible with accepted production methods. The spandrels of the Amex building in Brighton, England (preceding page), made of glass-reinforced polyester (GRP) are a notable exception. This componentized

plastic house near New Haven, Connecticut, designed by Yale architecture graduate Valerie Batorewicz is another rare example of exploiting these materials. Dismayed by the quality of much of the housing she had seen both in Europe and in the U.S., she traveled the country in the hope of discovering an up-to-date technology that could lead to new forms and more interesting spaces. When she saw what was possible with the fiberglass-reinforced plastic (FRP—U.S. terminology) used for auto bodies, Valerie determined that, for a new approach, "this was the only way to go."

Thus inspired she found financial support privately, a client, and a factory to fabricate the house shown here for a site near New Haven. Up until 1975, the cost of this construction, called "Environ A," for which she has a mechanical patent, was about $22 per sq ft, she says. The construction was 3½

in. of isocyanurate foam with 1/8 in. of FRP on the outside and 1/16 in. FRP inside, both given a mineral coating as a fire retardent that looks like plaster, all painted with acrylic latex. The bathroom/kitchen cores, the floors and the end panels of the FRP sections were framed in wood (the end panels were framed in wood in order to speed field construction). To obtain approval from the state, Spiegel & Zamecnik, structural engineers for the building, conducted load tests.

Because of higher petrochemical prices after 1975, a fabricator of FRP components, also in Connecticut, who built two dozen single-story houses with FRP, at a price competitive with "nicer homes," has been investigating the possibilities of fiberglass-reinforced concrete (FRC)—used for some time in England and increasing in acceptance in this country (see photos across page showing fabrication methods and two examples).

Cem-Fil Corporation photo

As the cost of resins has risen following the OPEC price jump in oil, interest has mounted in fiberglass-reinforced concrete (or glass-reinforced cement, as it is known in England, and to some extent here). Such a composite was not possible until an alkali-resistant glass fiber was developed by an English glass manufacturer. The building at top is a suburban office building of GRC in Nashville. The one at the bottom is the Lutheran Social Services Center in Minneapolis. Center photos show fabrication of FRC panels in molds at Concrete Technology, Inc., near Dayton, Ohio.

Sandwich construction for the ''plastic' house shown across page and below comprises a 3½-in. isocyanurate core for insulation with structural skins of fiberglass-reinforced plastic. The shell components were fabricated using molds of FRP braced by wooden wales. After the core was molded, the FRP layers were sprayed on.

Owens-Corning Fiberglas Corporation photos

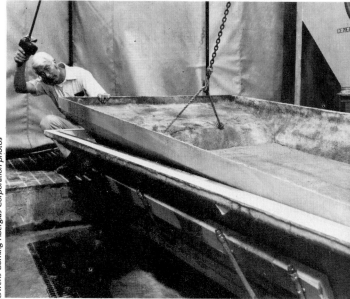

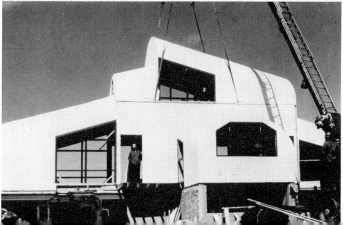

Complex concrete forms

The architects for this publicly owned garage in Charlotte chose architectural concrete as a finish material because it could serve as structure while also matching, at low cost, the buff-white facades of the adjacent county court house (limestone) and the facing county office building (precast concrete). Concrete also was a logical material for the helical drum that provides access to the second through fifth levels, and that, moreover, because of its turret shape and location on the landscaped courtyard, signals visitors to the court house where to park their cars.

According to Gerald Li, partner-in-charge for the architects, the expression of the exterior grew out of their massing scheme for the structure: a series of three steps providing a smaller scale on the street side and rising to the four-story height of the other two buildings on the courtyard. The designers wanted to expose the columns so the tiers would "read" as a series of table-like units. The remaining articulation of the facade followed in logical progression. Because it would have been difficult to butt spandrel and column forms, the fiberglass (FRP) column forms were provided with haunches. This established a vertical reveal. The rail panels atop the spandrels were stopped short of the columns to show they are not structural, and have a reveal matching that of the haunch. Spandrel and rail forms were divided into three sections between columns for ease of handling, establishing remaining reveals.

To ensure quality and to expedite construction, the architects engaged consultants Kelly/Hough, who not only developed a series of construction procedures and forming techniques, but, in a step unusual in construction circles, produced explanatory drawings to assist the contractor.

MECKLENBURG COUNTY PARKING STRUCTURE, Charlotte, North Carolina. Architects: *Clark Tribble Harris and Li, P.A.—Gerald Li, partner-in-charge; Roger Hinton, project designer; Thomas D. Byrum, job captain.* Engineers: *Clark Tribble Harris and Li (structural); Geotechnical Engineering Company (soils).* Consultants: *National Planning, Inc. (parking); Kelly/Hough, Inc. (concrete).* Contractor: *Hickory Construction Company.*

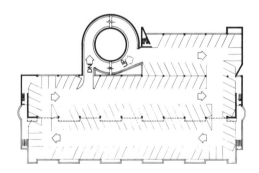

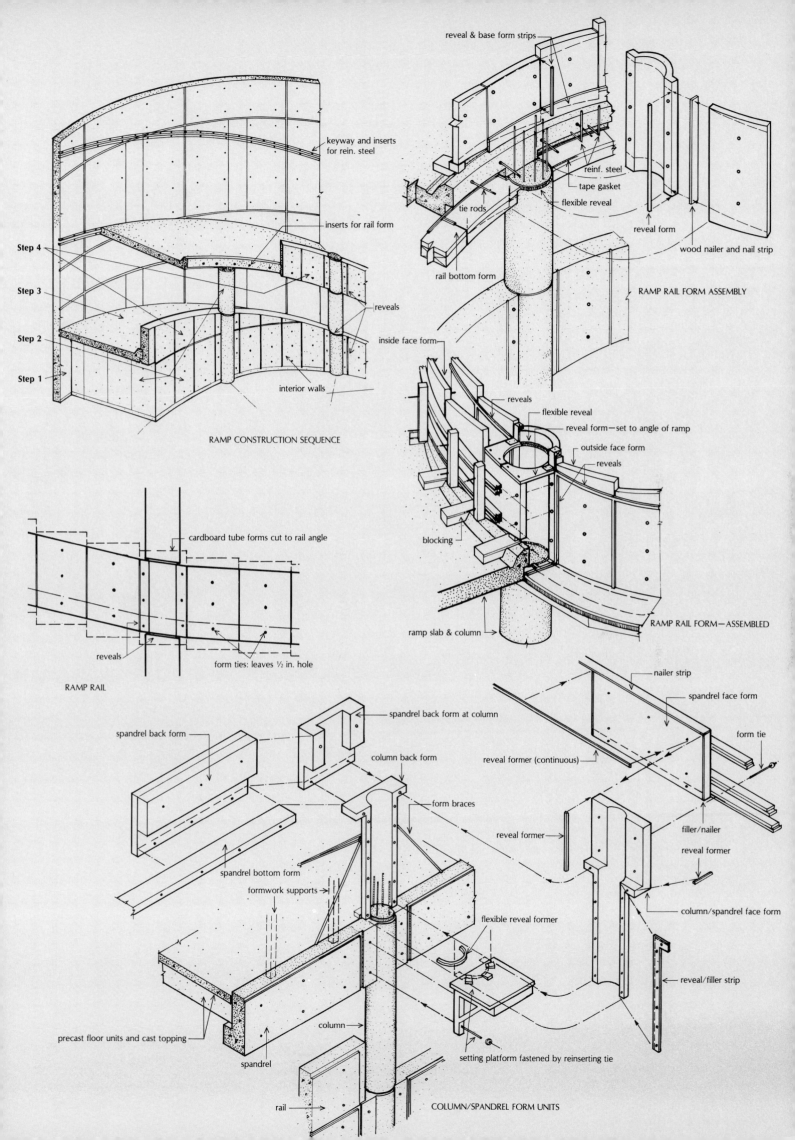

RAMP CONSTRUCTION SEQUENCE

keyway and inserts for rein. steel

inserts for rail form

Step 4

Step 3

Step 2

Step 1

reveals

inside face form

interior walls

reveal & base form strips

reinf. steel

tape gasket

flexible reveal

tie rods

rail bottom form

reveal form

wood nailer and nail strip

RAMP RAIL FORM ASSEMBLY

reveals

flexible reveal

reveal form—set to angle of ramp

outside face form

reveals

blocking

ramp slab & column

RAMP RAIL FORM—ASSEMBLED

cardboard tube forms cut to rail angle

reveals

form ties: leaves ½ in. hole

RAMP RAIL

spandrel back form at column

spandrel back form

column back form

form braces

reveal former (continuous)

reveal former

spandrel bottom form

formwork supports

precast floor units and cast topping

spandrel

column

rail

flexible reveal former

setting platform fastened by reinserting tie

nailer strip

spandrel face form

form tie

filler/nailer

reveal former

column/spandrel face form

reveal/filler strip

COLUMN/SPANDREL FORM UNITS

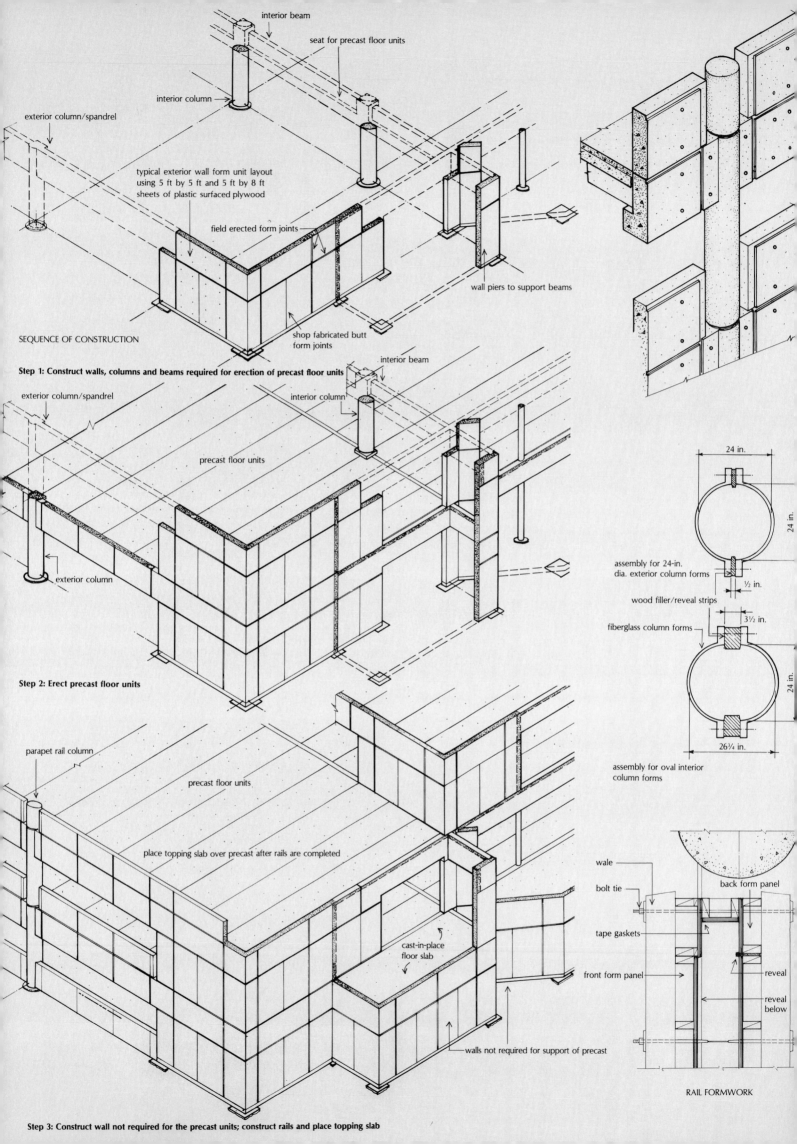

interior beam

seat for precast floor units

interior column

exterior column/spandrel

typical exterior wall form unit layout
using 5 ft by 5 ft and 5 ft by 8 ft
sheets of plastic surfaced plywood

field erected form joints

wall piers to support beams

SEQUENCE OF CONSTRUCTION

shop fabricated butt
form joints

Step 1: Construct walls, columns and beams required for erection of precast floor units

exterior column/spandrel

interior beam

interior column

precast floor units

exterior column

Step 2: Erect precast floor units

parapet rail column

precast floor units

place topping slab over precast after rails are completed

cast-in-place
floor slab

walls not required for support of precast

Step 3: Construct wall not required for the precast units; construct rails and place topping slab

24 in.

24 in.

assembly for 24-in.
dia. exterior column forms

½ in.

wood filler/reveal strips

3½ in.

fiberglass column forms

24 in.

26¾ in.

assembly for oval interior
column forms

wale

bolt tie

back form panel

tape gaskets

front form panel

reveal

reveal
below

RAIL FORMWORK

CHAPTER 5:

Mechanical Systems—New Concepts

Engineers have always been interested in saving energy. As long as 25 years ago, engineer Frank Bridgers designed a mechanical system for a New Mexico office building that takes the sun's heat from the south side and delivers it to a heat-pump system that redistributes the heat to the cold side in winter. And 20 years ago he conceived the idea of using large-size packaged air conditioners for lower initial cost in heating and cooling of schools.

When energy was relatively cheap, often more effort was put into saving initial costs than into saving energy. (One positive result of those times, however, was the development of factory packaging of more sophisticated equipment than had been factory assembled before, suitable for a wide range of building sizes and types.) Wasteful systems have been discarded; or their application has been modified to avoid waste. Air-handling systems have been broken down into smaller units than before, offering the cost-saving advantage of factory-built equipment and, in tall buildings, allowing each floor to have its own air-handling system—with the attractive benefit of being able to control each floor independently with respect to nighttime and week-end use, and with respect to smoke control in case of fire.

This chapter deals with some of the more innovative developments in mechanical systems in the last 10 years, particularly in packaging concepts and in heat pumps. Integration of mechanical systems and lighting will be taken up in the following chapter. And, finally, this chapter presents a series of architecturally significant buildings heated and cooled by using solar energy.

Energy saving in buildings can be accomplished in design via brute-force techniques or finese. The brute-force route is simple: use less lighting, less glass; reduce optimum temperature levels, etc. For the mechanical and electrical engineer, and the lighting designer, the finesse route means bringing to bear a greater depth and breadth of knowledge, putting in more time for investigation of alternatives for design (which the client should pay for), and embarking upon projects with some courage and conviction—and with the attitude that husbanding energy resources is the engineer's responsibility along with the architect's and the owner's.

Frank Bridgers of Bridgers & Paxton, consulting mechanical engineers of Albuquerque, has always had a deep interest and belief in energy-conserving systems—and his interest predates the start of his firm in 1951. Some of his interest undoubtedly stems from his graduate engineering work with F.W. Hutch-

Energy management approaches

Albuquerque, New Mexico, site of the Bridgers & Paxton office building, has a maximum of solar radiation in the winter—so it was an ideal situation to try out the concept. The building shape reflects an optimized angle for the collectors. The system includes storage and a heat pump for cooling as well as heating.

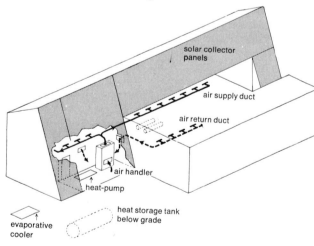

inson at Purdue University, who pioneered early mathematical analysis for radiant heating panel design; and also from his early experience in the consulting engineering firm of the late Charles S. Leopold of Philadelphia—who designed some of the earliest commercial installations of radiant-panel heating and radiant-panel cooling, and also the earliest test installations of water-cooled lighting fixtures.

Bridgers' concern with heating efficiency predates the energy crisis by 20 years

A reflection of Bridger's interest in energy conservation is the firm's own office building, designed in 1954 to use solar collectors for space heating. One whole side of the buidling was covered with solar collectors, and sloped for optimum solar energy recovery. A packaged water chiller, used for air-conditioning in summer is used as a heat pump during the heating season, combined as a system with the solar collectors and an underground storage tank. If heat from the sun is inadequate on cloudy or very cold days, heat is drawn from the stored water.

The building has been instrumented by Dr. Stanley Gilman of The Pennsylvania State University through a National Science Foundation grant, and data are being developed by Bridgers and Gilman to provide application guidance for other engineers.

Bridgers also was one of the first engineers in the country to design a heat-pump system for a high-rise office building—with the main objective being to pick up heat from one side of the building that required cooling and transferring it to another calling for heating.

Furthermore, the system included radiant heating-cooling panels in the perimeter ceiling and under the windows to reduce air-conditioning supply-air requirements—which was aided further by the novel sill air-return arrangement that helped minimize heat-transmission effect through the glass into the occupied space.

Since then, Bridgers has done a number of internal-source heat-pump designs, including the use of large unitary (packaged) water-cooled air conditioners in a school that pulled heat from the interior in cold weather and utilized it on the perimeter where heat is needed. Bridgers has shown that larger-size, heavy-duty package air conditioners up to 60 tons in refrigeration capacity may offer an attractive alternative to central chilled-water-plant systems. He has also done a number of buildings with large central-plant, internal-source heat pumps—giving special attention to some of the idiosyncracies of refrigeration-plant design that can cause operational difficulties—particularly with respect to the method of heat rejection from the chiller's condenser.

1954

In the Simms Building in Albuquerque, Bridgers designed an air-return system that captures some of the sun's heat and delivers it to a heat-pump system that redistributes the heat in winter to the cold side. Radiant heating-cooling panels, tied in with a solar compensator control, reduce air-conditioning air-supply requirements. The panels provide perimeter heat on the north side in winter, and on the south side when there is no sun. Well water is available as a standby if there is not enough solar energy or internal heat.

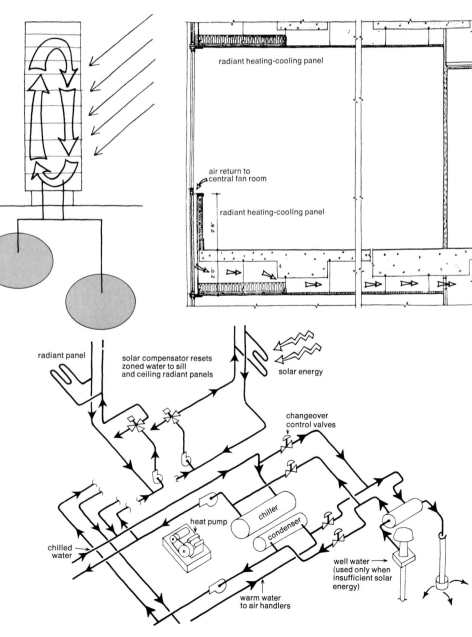

radiant heating-cooling panel

air return to central fan room

radiant heating-cooling panel

radiant panel

solar compensator resets zoned water to sill and ceiling radiant panels

solar energy

changeover control valves

heat pump

chiller

condenser

chilled water

well water → (used only when insufficient solar energy)

warm water to air handlers

Benefits from solar energy are possible in both ordinary and special buildings

Frank Bridgers looks at the beneficial utilization of solar energy from two different standpoints: 1) use of solar collectors to pick up heat directly or indirectly from the sun's rays; and 2) controlling the sun's heat that manages to penetrate the building's glass. He was recently involved in the design of one of the largest solar-collecting systems for space heating application yet to be used—Denver Community College. One of the practical problems he had to deal with was how to prevent the freezing of the collectors at night. The solution was to provide an anti-freeze loop for the collector circuit only, with a heat exchanger to transfer heat to the solar storage tank. Although this solution reduces the efficiency of heat transfer to some extent, the operation and maintenance problem of venting air from a large amount of collection piping everyday in winter would be a less practical solution. Bridgers points out that the technology of solar collection has changed little in 25 years; but, on the other hand, engineering data for design, the collectors themselves, and economic feasibility are changing.

Beneficial utilization of transmitted solar energy has been an integral part of the mechanical-system design of several office buildings Bridgers has been involved with. What has to be stressed in the future, he says, is a proper concern with building orientation, combined with carefully chosen (or designed) interior and exterior shading or sun control to make it possible not just to limit or reject solar radiation in summer, but also to "capture" it in wintertime.

The more dedicated engineers, such as Frank Bridgers, continually strive to improve upon their past designs, to broaden their background expertise, and to investigate new approaches involving a wider range of design solutions. Furthermore, they put together highly-qualified engineering "teams" that get in-depth direction from a highly-qualified principal of the firm. Characteristically, the principals and chief associates of such firms have high educational qualifications—formal or otherwise—combined with broad practical design, installation, and field-testing expertise on a wide range of systems and equipment.

Bridgers has considered active technical and administrative participation in the American Society of Heating, Refrigeration and Air Conditioning Engineers—the consulting engineers' primary technical society—to be an important responsibility. Furthermore, his educational-engineering background has made him particularly aware of the value of basic fundamental data developed by ASHRAE. Designing a system that is *practical* is always uppermost in Frank Bridgers' approach, and some of the specific elements of that approach follow.

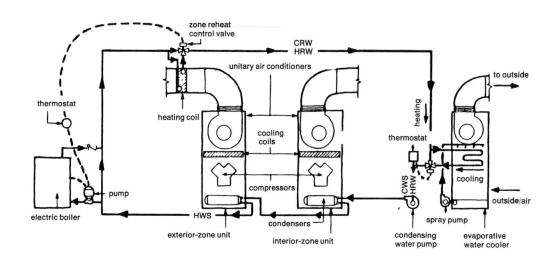

1968

Bridgers conceived the idea of using packaged unitary air conditioners for both heating and cooling in low-rise buildings with large interior areas. The interior units in cooling pick up heat and transfer it to exterior units. A boiler makes up any deficit in heat, and an evaporative water cooler rejects excess heat.
The approach is best for buildings with adequate space for low-pressure ductwork.

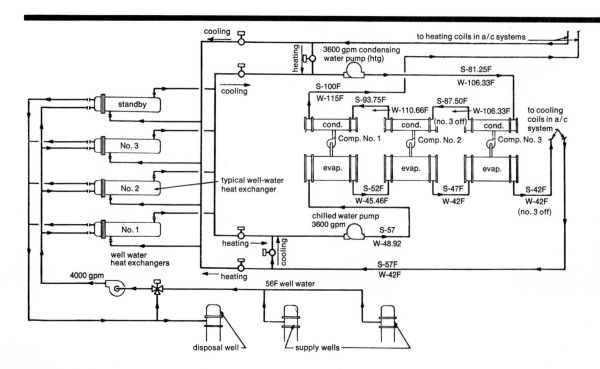

1972

This large (2500-ton) heat-pump system for the huge, high-rise LDS office tower in Salt Lake City uses well water as a source. At times internal heat will balance external load of the building. When more heat is required it can get it from the well water. If, on the other hand, there is excess heat, it is rejected to the well water. Use of separate heat exchangers avoids potential corrosion problems from well water.

Innovation can be made practical by using standard equipment packages

Bridgers' guideline to design for practical innovation include:

1) Use standard production-line equipment packages, combined with manufacturers' standard options—as proven by field experience. The component selection and arrangement and the piping design in a refrigeration system are critical. Use of other than standard "matched-performance-package" configurations can be troublesome and costly, and can limit bidding. This caution, to some extent, applies to centrifugal chiller heat pump systems employing double-bundle condensers because of the complexities in evaluations for stability of operation at the lower heating-load conditions they encounter;

2) Don't ask for custom features with which the manufacturer has not had experience. Rather, analyze the detailed performance of standard heating-cooling packages to see how they might be used in new and unique systems to improve over-all efficiencies;

3) When applying standard packages in new systems, think through the over-all system including: a) pipe and duct system dynamics, b) heat-exchange options and limiting factors, c) selection of control-system elements and their over-all coordination—perhaps the most essential element to ensure that the hvac system will perform as designed;

4) With unique systems, make sure the contractor is guided in terms of start-up, balancing, testing, and adjustment of the system. This means the consulting engineer must have experienced and knowledgeable engineers in these areas;

5) Optimize duct and piping system design and fan selection to achieve an optimum balance between first cost and operating cost of the heating and air conditioning system.

1978

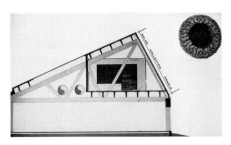

Denver Community College has one of the largest solar heating systems in the country. The architects, ABR Partnership of Denver, favored the approach because it is pollution-free, would not be affected by the shortage or cost of natural gas, and because the climate of Denver is ideal for a solar system. There are 35,000 square feet of collector surface tilted at 53 degrees from the horizontal. Solar-heated water is stored in 200,000-gallon-capacity tanks underground. If the sun heats it to over 100 F it is used directly, if not, a heat pump system adds supplementary heat to raise it to 100 F.

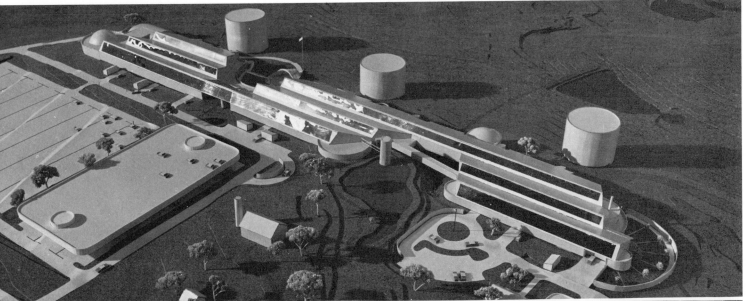

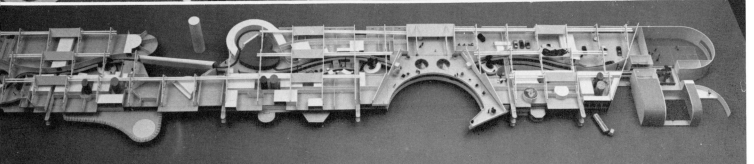

Individual packaged air handlers

It is not heresy, but rather a series of practical considerations that are involved in some forward-looking engineers' use of individual air-handling systems on every floor of some recent high-rise office buildings, instead of centralized fan rooms girdling a building every 15-20 floors.

One of the most cogent reasons for this approach is life safety: smoke and lethal gases cannot be spread by the fan system to other floors. But, beyond this, the initial cost and operating economics are very favorable.

Engineers say the air-handling apparatus and ventilation shafts can fit in the same space or less than was formerly required for the air shafting alone.

Another advantage on the operating side is that people working overtime can be provided with heating or air conditioning at a reasonable cost because fans can be operated individually.

A Robert T. Tamblyn, Toronto consulting engineer, has developed a "compartment" concept in which one fan serves an entire floor, both exterior and interior. The system is single duct with variable-air-volume terminals to provide temperature compensation for various zones and exposures. Tamblyn saves money by using all overhead air supply; so, to take care of building skin heat loss in winter, he calls for shallow (only 3 to 6 in.) fin-tube radiation at the perimeter of the building.

The system calls for variable air-volume fan systems stacked in vertical alignment so that they can be served from one shaft for the preconditioned ventilation air.

Special attention was paid to the design of the "compartment" fan units to get top performance in a minimum of space. They were developed over a two-year period, in consultation with several manufacturers, for precise acoustical, dynamic and thermal performance. Further, high standards were adhered to in the selection of materials and components.

The Canadian National Building Code makes it mandatory for all new high-rise buildings to have a masonry smoke shaft, unless the building is fully sprinkled.

There are other approaches such as providing axial-type ventilation fans that can be reversed in the event of fire, with an elevator shaft being dampered and opened to relieve the smoke and gasses.

B A different approach to the compartment concept was developed by another Toronto firm, G. Granek and Associates, for the 20-story Guardian Royal Exchange Tower. It uses fan units on each floor for the interior zone, but has a conventional perimeter induction-unit system for the exterior zone. This induction-unit system supplies the makeup ventilation air for the whole floor. The building also has a separate masonry-enclosed smoke shaft for emergency automatic operation. The interior-zone system only recirculates and cools the air; return is via the hung-ceiling plenum.

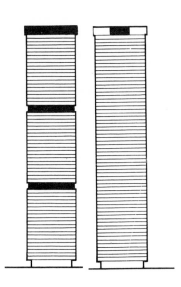

Architects: Crang and Boake
Engineers: Tamblyn, Mitchell and Partners, Ltd. (mechanical)

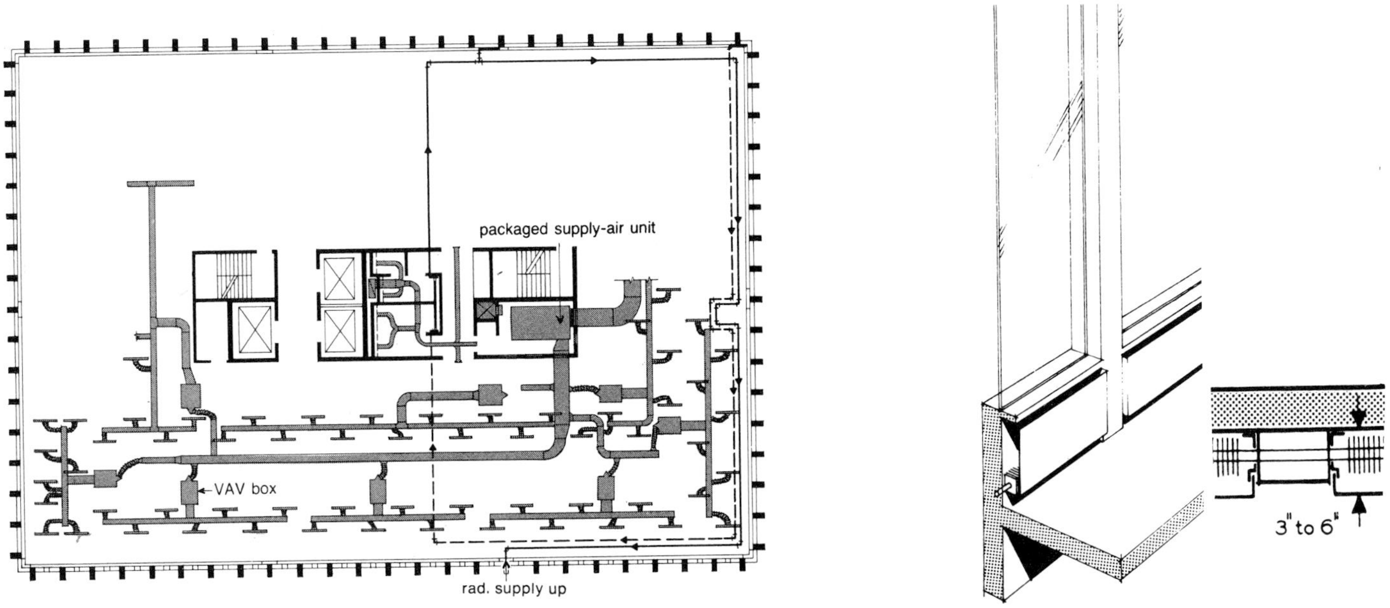

A

packaged supply-air unit

VAV box

rad. supply up

3" to 6"

Architects: Searle Wilbee Rowland
Engineers: G. Granek and Associates, Ltd. (mechanical)

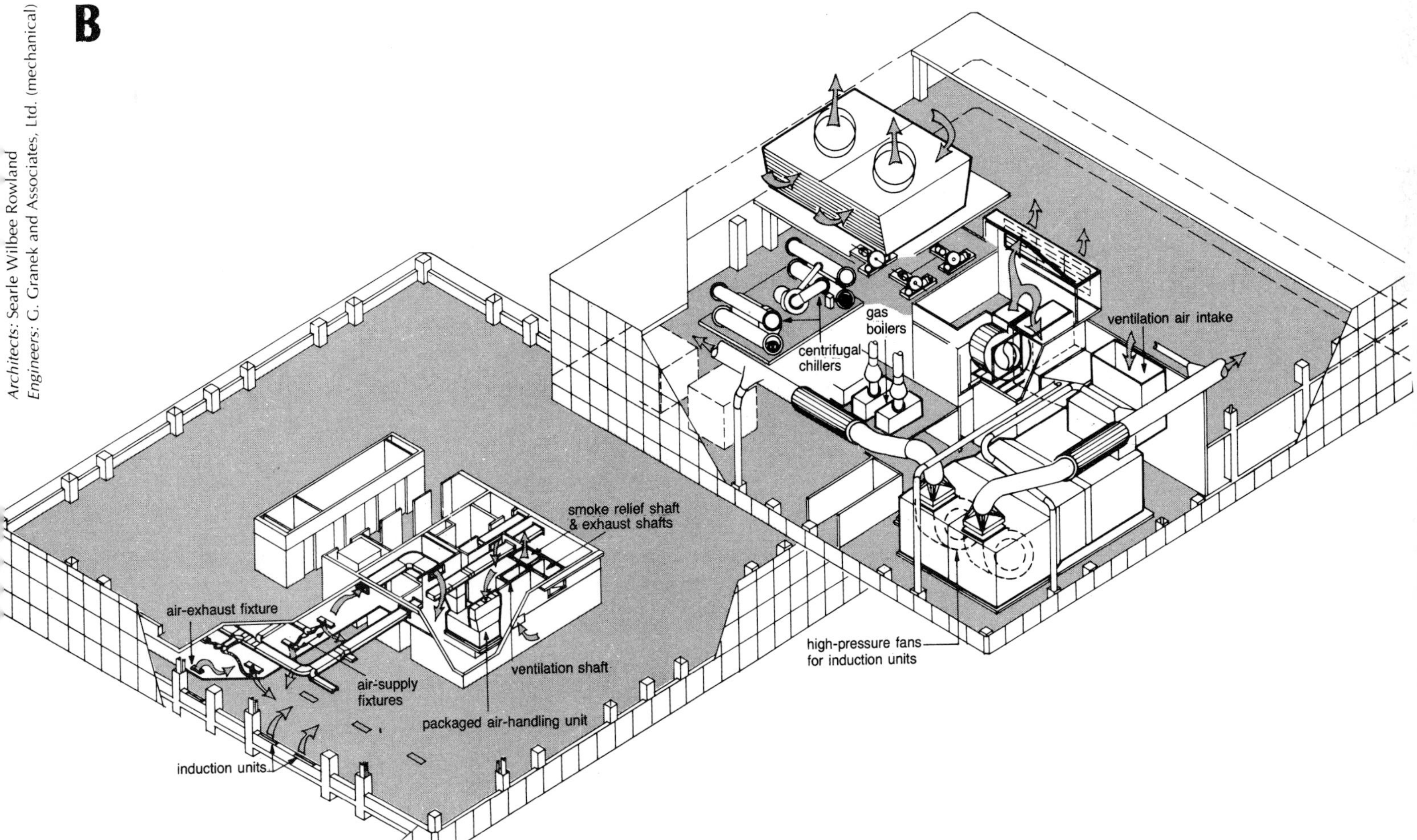

B

smoke relief shaft & exhaust shafts

ventilation air intake

gas boilers

centrifugal chillers

air-exhaust fixture

air-supply fixtures

ventilation shaft

packaged air-handling unit

high-pressure fans for induction units

induction units

Small single-package heat pumps

Providing thermal comfort conditions in modern buildings often has turned out to be, thermodynamically, redistributing heat. Thus, we have seen the frequent application of what have been termed "internal source heat pumps" with large central systems—recouping heat from lights, people, and sometimes the sun, and rerouting it to spaces that need heat. Comparably, on the unitary side there is the water-to-air unitary heat pump, in sizes of ¾ to 5 tons, that is tied into a water loop for either abstracting or rejecting heat. Interior units generally need to produce cooling, except for morning warm-up in winter. When cooling, they reject their heat into the loop which is taken out in winter by the perimeter units that are calling for heat. In summer, when all units are cooling, the heat is rejected to the atmosphere via a closed-circuit evaporative cooler. If more heat is needed in winter than is rejected by interior units it usually is supplied by an electric boiler, or from a storage tank having electric immersion heaters.

The two installations shown here are suburban office buildings whose clients wanted quality architectural design as well as quality hvac at reasonable cost.

A The Geico office building was conceived as an all-electric building because of the shortage of gas for heating in the Long Island area, and because of the wish to avoid the cost and environmental complications of an oil-fired boiler plant. The heat pumps are all ceiling-mounted, except for the cafeteria and telephone room.

The basic open floor plan had to be considered in mechanical system selection. The use of ceiling-hung units allowed maximum utilization of floor space, and also gives the owner maximum flexibility when rearranging offices.

Each heat pump is controlled by a space thermostat, and each has automatic change-over controls to ensure prompt reaction to space conditions.

The ceiling space also is used as a return-air plenum, minimizing the amount of ductwork required. Return air is through heat-recovery light troffers, directly picking up heat from the lamps—reducing heating load to the space, and lowering operating temperature of the lamps which adds to their life.

A simple heat-recovery wheel system for ventilation air conditions is used so as to reduce heating and cooling loads. Condenser water storage minimizes electric-load peak of the system.

B The Volvo building has long-span, column-free, open-plan spaces, whose deep girders made it possible for the engineer to consider above-ceiling heat pumps in the 5-ton size, which are light-weight and vibration-free. At the perimeter, on the other hand, console units are provided, using architect-designed custom enclosures. Minimum ventilation is provided through a separate outside air system ducted to the ceiling heat pumps. Floor-standing units are in the smaller wing.

Lawrence S. Williams, Inc. photos

Bill Rothschild photos

A

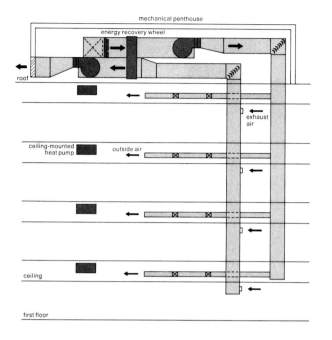

Architects: Vincent G. Kling & Partners. Engineers: Kling-Lindquist (mechanical/electrical)

mechanical penthouse
energy recovery wheel

roof

exhaust air

ceiling-mounted heat pump
outside air

ceiling

first floor

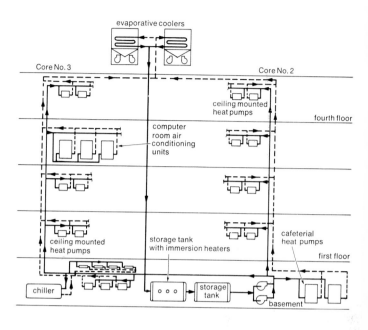

evaporative coolers

Core No. 3
Core No. 2

ceiling mounted heat pumps

fourth floor

computer room air conditioning units

ceiling mounted heat pumps
storage tank with immersion heaters
cafeterial heat pumps

first floor

chiller
storage tank
basement

B

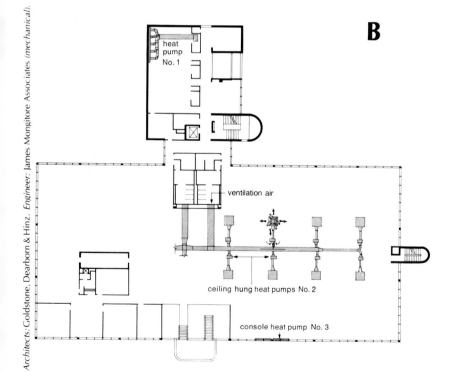

Architects: Goldstone, Dearborn & Hinz. Engineer: James Mongitore Associates (mechanical).

heat pump No. 1

ventilation air

ceiling hung heat pumps No. 2

console heat pump No. 3

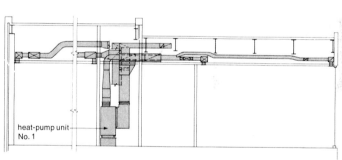

heat-pump unit No. 1

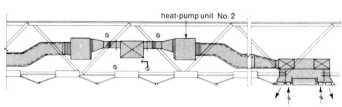

heat-pump unit No. 2

Multizone air handlers for a high-rise office building

The developers of this building put together a team of specialists in design, leasing, engineering, and construction to meet the challenges of difficult site, budget, construction time, and life-cycle costs.

Mechanical and electrical systems were budgeted at a low figure of $5.33 per sq ft (1974), demanding close attention to hvac-system-related costs. The mechanical/electrical engineers estimate that use of reflective, insulating glass, for example, will save $36,000 per year, net, in capital and operating costs.

The hvac system saves energy in three ways: First, the two 1,200-ton chillers have double-bundle condensers that capture internal heat for reuse at the perimeter. Second, each floor has two multizone-type fan-coil air handlers, each designed for eight zones—several serving the perimeter, and the balance, interior. Thus, each floor can be operated separately. Third, the air-handling system never "reheats" already cooled air, as some multizone systems do. In the heating mode, cooling-coil face dampers are fully closed and bypass dampers are fully open, so no air is blown through the cooling coil. A thermostat-controlled, three-way valve modulates supply of hot water to the heating coils. Water is distributed by a four-pipe system.

PEACHTREE SUMMIT, Atlanta. Owners-Developers. *Diamond & Kay Properties and P.C. Associates.* Architects: *Toombs, Amisano & Wells.* Engineers: *Ellisor Engineers, Inc.* (structural); *Herman Blum Consulting Engineers* (mechanical/electrical). Contractors: *Henry C. Beck Company* (general); *Sam P. Wallace Company, Inc.* (mechanical); *Fischbach & Moore* (electrical).

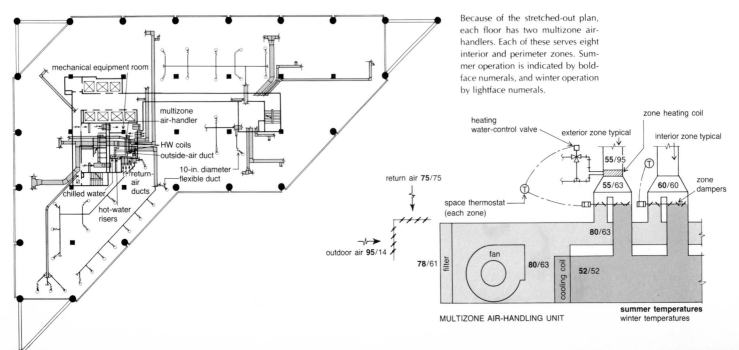

Because of the stretched-out plan, each floor has two multizone air-handlers. Each of these serves eight interior and perimeter zones. Summer operation is indicated by boldface numerals, and winter operation by lightface numerals.

mechanical equipment room

multizone air-handler

HW coils outside-air duct

10-in. diameter flexible duct

return-air ducts

chilled water

hot-water risers

heating water-control valve

exterior zone typical

zone heating coil

interior zone typical

55/95

55/63

60/60

zone dampers

space thermostat (each zone)

80/63

return air 75/75

outdoor air 95/14

78/61

filter

fan

80/63

cooling coil

52/52

80/63

MULTIZONE AIR-HANDLING UNIT

summer temperatures
winter temperatures

All-air system for a skyscraper

All-air systems of the double-duct type have been popular in Los Angeles because the mild climate has been in their favor, and because fewer trades are involved in their installation.

But in a tall structure such as the 62-story United California Bank Building, shaft space of a double-duct system could eat up a lot of valuable rental area.

To minimize this problem as well as to cut down fan room height at the 22nd- and 42nd-floor levels (there are fan rooms at levels 4, 5, 22, 42, 61 and 62),

the engineers, the Los Angeles office of Syska & Hennessy, designed a unique variable-air-volume system that only requires single-duct risers, cutting down on shaft space required. In addition, fan room height can be lower because only one set of coils is required rather than both hot- and cold-deck coils.

Each floor has two duct loops, one cold and the other warm. The cold loop supplies single-duct variable-air-volume (VAV) terminals in the interior space, and both loops serve var-

iable-temperature, constant-volume double-duct boxes serving the perimeter offices.

Altogether there are five air-handling systems. Three of them are cooling only—feeding the inner loop. System No. 4, however, serves either as a cold-air supply, or, in cold weather, as the main hot-air supply. This change-over concept necessitated the use of continuous hot- and cold-loop ducts, with switchover provisions for System No. 4. Because some heating is required in intermediate-season, sunny weather,

when System No. 4 is still needed for cooling duty, System No. 5 (a small, recirculation-only system) is provided for heating-only duty.

Return-air lighting fixtures are used that draw air through the lamp chambers. The fan systems conserve heat energy during the cooler weather: Fan System No. 4 is set for minimum ventilation to get maximum benefit from the warm return air, while the three cooling-only systems operate with increased outdoor air.

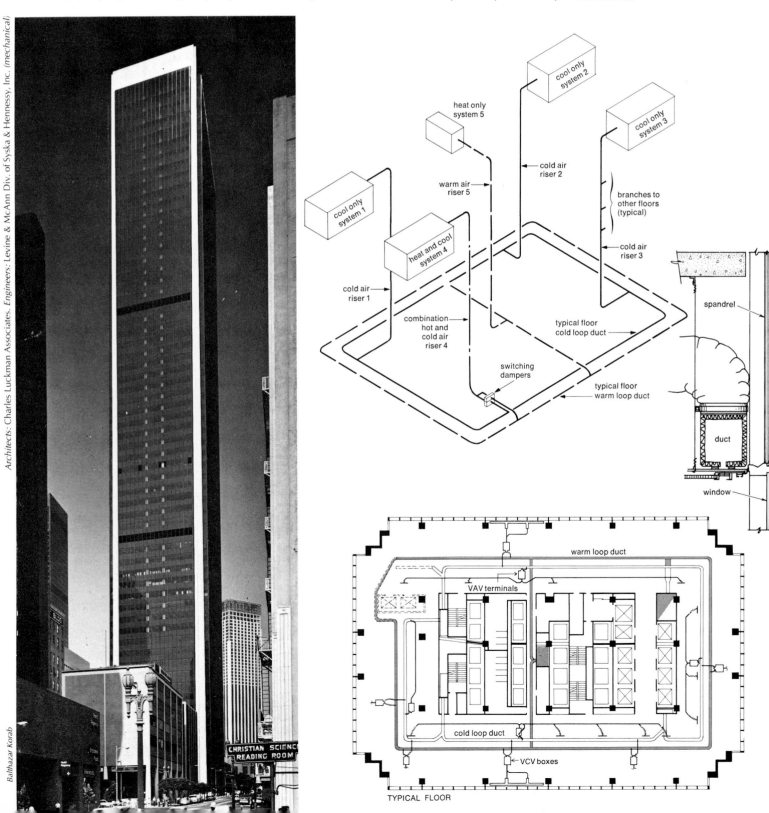

CHRISTIAN SCIENCE READING ROOM

cool only system 2

heat only system 5

cool only system 3

cold air riser 2

warm air riser 5

branches to other floors (typical)

cool only system 1

cold air riser 3

heat and cool system 4

cold air riser 1

spandrel

combination hot and cold air riser 4

typical floor cold loop duct

duct

switching dampers

typical floor warm loop duct

window

warm loop duct

VAV terminals

cold loop duct

VCV boxes

TYPICAL FLOOR

Multiple package heat pumps for a school

In Westminster, Colorado, a suburb of Denver, the most economical fuel is natural gas, despite a local shortage that led the utility to ration its supply for the new high school to 7,500 cfh, while similar buildings in the area use about 17,500 cfh. Attacking this problem, architects and engineers determined to minimize wastage, using insulation and insulated glass, and to make the most of everything by redistributing air to other spaces and recovering energy from exhaust.

The owner's requirement that most parts of the buildings be air-conditioned suggested the use of a water-loop heat-pump system, which can either heat or cool, as well as redistribute heat from warmer to cooler areas. Further, conditioned air that would normally be exhausted locally is circulated to other warmer (or cooler) areas by a fan system, limiting the need for fresh air.

The basic components of the system include 76 water-to-air unitary heat pumps, ranging in capacity from 1 to 5 tons, and a gas-fired boiler, which provides hot water both for the loop and for domestic use. Because of the efficiency of the heat-pump system, the superior insulation of the building shell, and minimal intake of fresh air, boiler requirements for gas were held to 2,700 cfh, 40 per cent of the consumption in conventional systems, say the engineers.

Supplemental domestic water heating is provided by an electric immersion heater, needed only at times of peak demand for both space heating and domestic hot water. In extreme cold, heat can be extracted from the 2,000-gal storage tank. If the gas supply is temporarily curtailed, the immersion heater can provide some space heating via the water loop.

In the school's shops, a heat recovery unit salvages heat from noxious exhaust, and filters remove dust from recirculating air.

WESTMINSTER HIGH SCHOOL, School District #50, Westminster, Colorado. Architects: *William Blurock & Partners.* Engineers: *Martin & Tranbarger* (structural); *Nack & Sunderland* (mechanical); *Frederick Brown Associates* (electrical); *KKBNA* (civil). Contractors: *Construction Management Services Division, Mead & Mount Construction Co.* (general); *Howard Electric and Mechanical Co.* (mechanical/electrical).

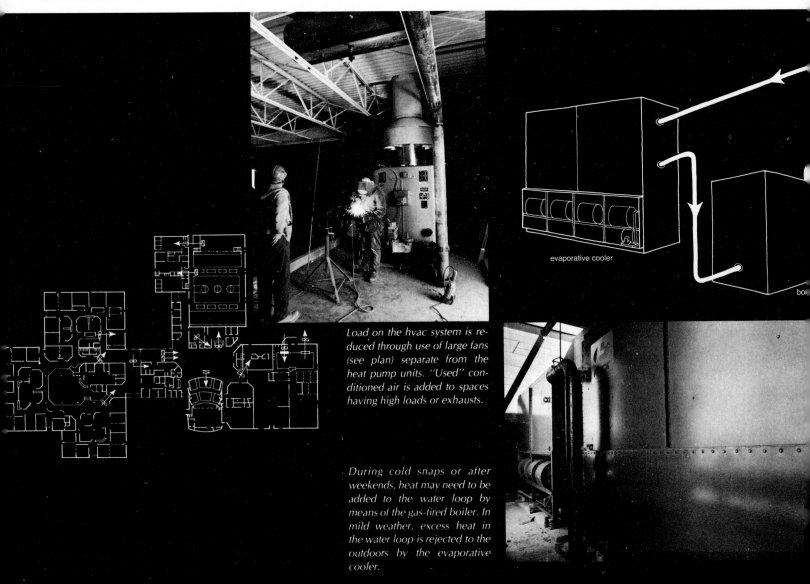

evaporative cooler

bo...

Load on the hvac system is reduced through use of large fans (see plan) separate from the heat pump units. "Used" conditioned air is added to spaces having high loads or exhausts.

During cold snaps or after weekends, heat may need to be added to the water loop by means of the gas-fired boiler. In mild weather, excess heat in the water loop is rejected to the outdoors by the evaporative cooler.

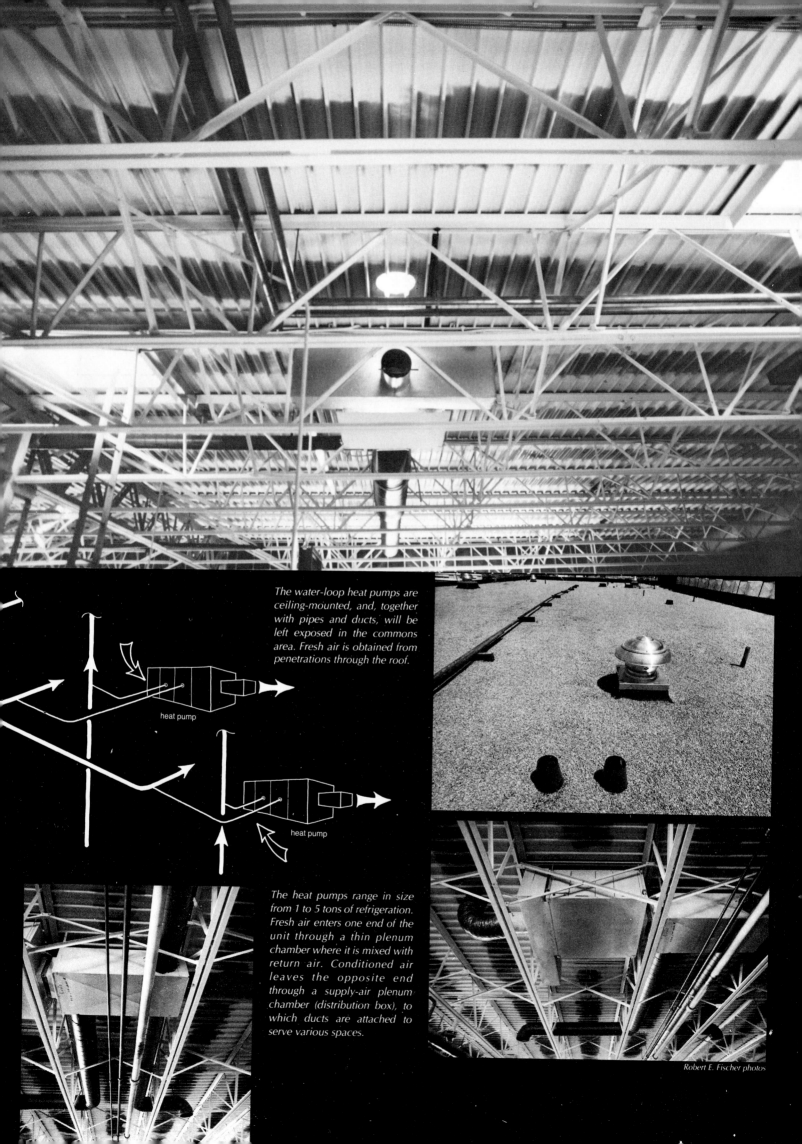

The water-loop heat pumps are ceiling-mounted, and, together with pipes and ducts, will be left exposed in the commons area. Fresh air is obtained from penetrations through the roof.

heat pump

heat pump

The heat pumps range in size from 1 to 5 tons of refrigeration. Fresh air enters one end of the unit through a thin plenum chamber where it is mixed with return air. Conditioned air leaves the opposite end through a supply-air plenum chamber (distribution box), to which ducts are attached to serve various spaces.

Robert E. Fischer photos

One of 76 water-loop heat pumps that was used in the Westminister School near Denver to reduce gas consumption for heating to 40 per cent of that normally used by conventional hot-water space heating systems. The ceilings were left open in the commons area of the school so the ceiling and joists were painted white, long-span trusses red, and circular ducts blue—to add a dash of color.

Heat pump with ice storage

A new strategy for reducing energy costs in the heating and cooling of residential, institutional and commercial buildings that has recently emerged from ERDA's Oak Ridge National Laboratory utilizes ice-making to get Btu's for heating in the winter and "free" cooling in the summer. The approach, called Annual Cycle Energy System (ACES), takes advantage of the physical fact that 144 Btu's of heat must be extracted from a pound of 32 F water to convert it to ice, and the same amount of heat must be applied to melt the ice. A crude parallel is old-timers' use of winter-frozen ice to preserve fresh foods in the summer.

At the heart of the system is a heat pump that extracts heat from the outdoor air during the heating season until the air's temperature drops to around 40 F. When the outdoor air is colder than this it "frosts up" the outdoor coils of the heat pump with a resulting drop in the efficiency. But what if the heat pump were to make ice in a large tank when outdoor air temperatures are at freezing or below? Not only would the efficiency of the heating cycle be improved, but the ice could be built up in the tank to be melted in warm weather.

This basic idea of making ice to heat a house was suggested by engineers over 40 years ago. Credit for reviving it belongs to engineer Harry C. Fischer who, as a heat-pump engineer in the '50s, tried to interest the industry in a heat pump that made ice for heating and kept the ice accumulated for space cooling. As Fischer notes, the idea didn't "fly" because of the cost of storage for the ice and the low cost of energy. But eight months after the Arab oil embargo, Harry Fischer, then in retirement, presented his ice-storage idea to the Energy Division at Oak Ridge National Laboratory (ORNL), which had been looking for conservation approaches for building thermal envelopes, water heaters and heat pumps. ORNL, operated by Union Carbide Corporation, liked the idea and hired him as a staff consultant to pursue development. In December 1974 HUD awarded Fischer's group a $100,000 grant for the ACES project, and by February, the first test ACES system was built and operating. A month later Westinghouse Electric Corporation, which had been urged by Fischer to build an ACES system, had one installed in their domestic engineering center near Pittsburgh. Now an ACES demonstration residence (shown right) has been built in Knoxville, Tennessee as one of several houses in the Tennessee Energy Conservation in Housing program sponsored by

This ACES (Annual Cycle Energy System) house at the University of Tennessee will have its operation compared with a solar house and a conventional house. The system comprises a heat pump on the second level behind the outdoor radiant/convector coil and a 16- by 21- by 8-ft ice bin. System makes ice in winter to get heat and melts it in the summer to get cooling. To make more ice in summer, refrigerant is rejected to the outdoor coil.

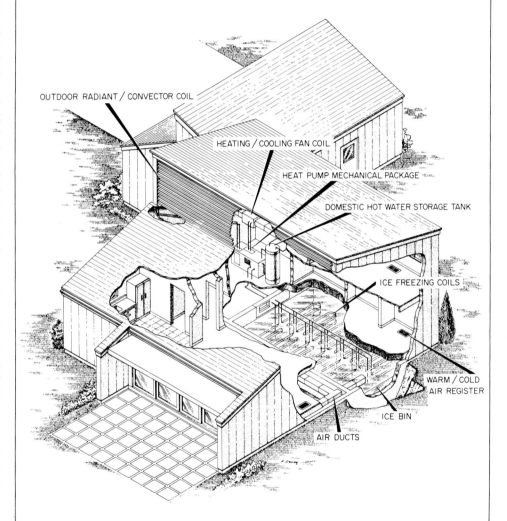

OUTDOOR RADIANT / CONVECTOR COIL

HEATING / COOLING FAN COIL

HEAT PUMP MECHANICAL PACKAGE

DOMESTIC HOT WATER STORAGE TANK

ICE FREEZING COILS

WARM / COLD AIR REGISTER

ICE BIN

AIR DUCTS

The ice bin has 1300 ft of aluminum finned tubing for its cooling coils. The ½-in. tubing and 3-in. fins are extruded as one section. Same material is used for the outdoor coil. In Knoxville, the ice bin has enough capacity to keep adding ice for the entire heating season, so there is no need for heat input from the outdoor coil. Ice made in winter will last most of the summer for cooling. When the ice has melted, the compressor will run at night to chill the tank's water.

ORNL, the University of Tennessee, and the Tennessee Valley Authority.

Cost of an ACES system for an 1800-sq-ft house has been estimated to be $5,000-6,000 for Washington, D.C. and Philadelphia, compared with $1,500 for electric heating and air conditioning (1976). Even so, engineers say, this is less than the cost for a completely solar heating-and-cooling system.

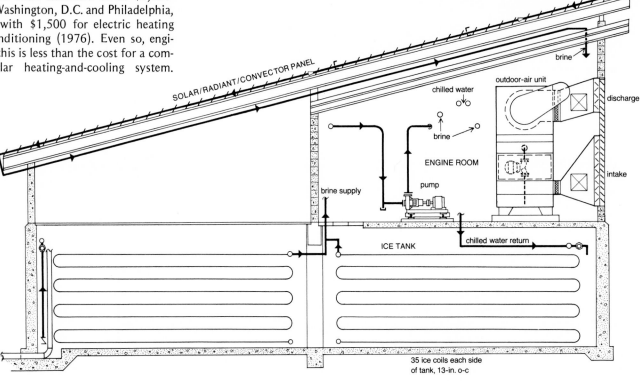

35 ice coils each side of tank, 13-in. o-c

And Oak Ridge engineers have been talking to manufacturers about application of ice-makers normally used for food preservation to reduce ice-tank costs: an ice-maker installed over a bin could merely slough off ice into the bin, eliminating pipe coils in the bin, itself. But until this is tried out in practice, some engineers will have questions about the efficacy of heat transfer from ice to water.

Because of the higher capital costs for ACES, its advocates are looking for applications in building types where owners are receptive to life-cycle costing. For this reason, they are eyeing a research installation by the Veterans Administration in a 60-bed nursing home at Wilmington, Delaware (illustrations this page). Though the nursing home had been designed when VA mechanical engineers heard about ACES, in their effort to promote energy savings they decided to have the project bid both conventionally and ACES, and brought in consulting engineer Robert G. Werden, a consultant to Oak Ridge on commercial development of ACES, to design the system. Werden—who has a long-time background in the development and application of heat pumps in commercial, industrial, school and apartment construction—designed an "energy bank" that will provide 75 tons of cooling and 800,000 Btu of heating. Because of the storage capability of the system, not only are operating-cost economics of the heat pump improved, but electric demand can be reduced greatly by making ice at nights during the summer. A special computer monitoring the system will decide which of eight modes of operation (shown across page) will satisfy space-conditioning demands, will control storage according to season, and will choose the most economical mode as governed by both weather and season.

Because the 60-bed VA nursing home in Wilmington had been designed before VA engineers decided to use the Annual Cycle Energy System, consulting engineer Robert Werden put his "energy bank" in a separate structure. It incorporates refrigeration (heat pump) equipment, a 40- by 50- by 10-ft ice tank, a solar/radiant/convector panel, and a computer system to monitor and control the operation. During the heating season, the outdoor air unit extracts heat from the air until it drops into the 40's. At lower air temperatures, refrigerant circulates through the brine (methanol) cooler so it can make ice. The roof panel absorbs solar energy to melt excess ice in winter, and rejects heat to outdoor air and the night sky in summer.

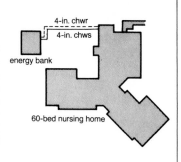

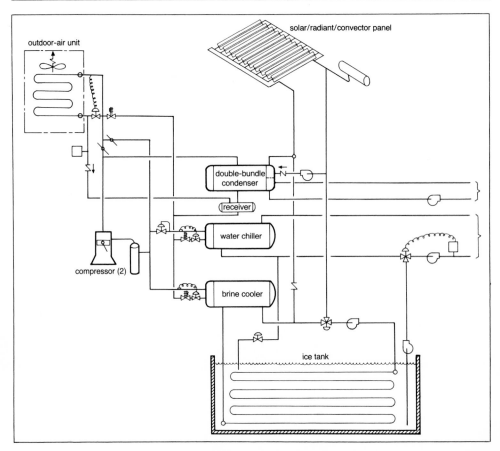

158

During the heating season, the outdoor unit extracts heat from outdoor air until the weather is in the 40's (Mode 1). In its least efficient operation, this cycle has a coefficient of performance (COP) of 4.6. Cold refrigerant circulates through the outdoor unit picking up heat, and the heat of compression from the two 37.5-ton reciprocating compressors (800,000 Btu/hr max.) is rejected to the double-bundle condenser for space heating.

When outdoor air is too cold for efficient operation of the outdoor unit, the system makes ice (Mode 2). Refrigerant chills the brine (methanol solution) to water-freezing temperatures, and the heat of compression, again, is used for space heating. Ice can build up on the coils to a 6-in.-dia. "doughnut," and, in this least-efficient condition, the COP is 3.38.

When there is useful solar heat and the system is calling for heating (Mode 3), brine is circulated through the solar panel and then through the tank to melt ice. Heat from this process is added to the heat of compression of the refrigerant.

If there is more heat available from the outdoor unit than is needed for space heating by itself (Mode 4) the excess can be used to melt ice in the tank, in preparation for more freezing later.

When outdoor temperature is mild during the heating season, and the building is not calling for heat (Mode 5), solar heat can be used to melt ice in the tank, in preparation for more freezing.

At the beginning of the cooling season (Mode 6), enough ice has been allowed to build up on the coils to provide cooling for the building without any need for running the compressor.

During the warmest weather when the ice has run out (Mode 7), the cooling effect can be produced by discharging the heat of compression to the outdoor unit, and also to the radiant/convector panel if the temperature of the brine in the panel is less than that of the refrigerant in the condenser.

When nights are cool in summer (Mode 8), the system can make ice off-peak for cooling during the warm day, saving energy and taking demand off the utility.

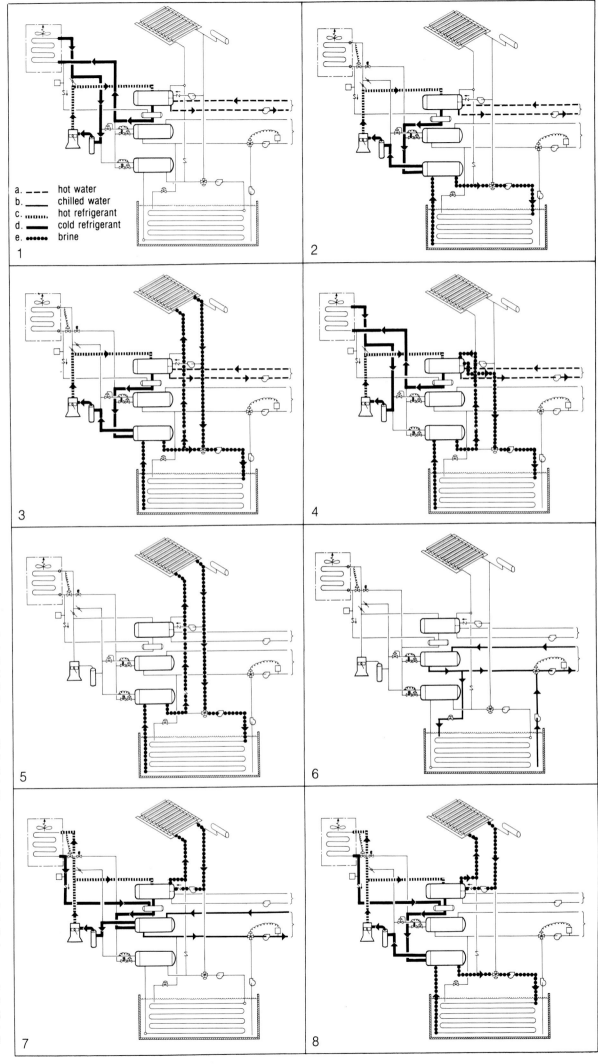

a. --- hot water
b. —— chilled water
c. ⋯⋯ hot refrigerant
d. ▬▬ cold refrigerant
e. •••• brine

Reginald Wade Richey

It is clear from the examples on pages 160 to 177 following, that architects are beginning to take solar seriously. Though paybacks on collector systems may not exactly thrill business clients, some systems do offer quite respectable paybacks. In any case, we should all be thankful to those clients, both public and private, willing to take the plunge and allow the rest of us to learn from their design and operating experience. Architects for some of the buildings shown here found that, with energy-saving techniques and economical building materials, total costs could be made comparable to those of like buildings.

But just as important, other examples demonstrate some inventive ideas for the use of solar heat and light through windows. Architects must recognize that this "passive" approach can lead to good design—and that more and more clients will be asking for it.

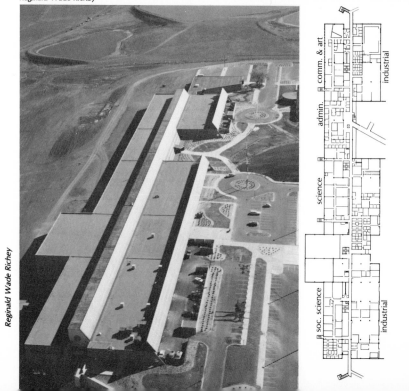

Reginald Wade Richey

Solar-heated campus: an active solar system

COMMUNITY COLLEGE OF DENVER-NORTH CAMPUS, Denver, Colo rado. Architects: *John D. Anderson and Associates.* Engineers: *KKBNA, Inc. (structural); Bridgers and Paxton (mechanical, energy conservation and solar energy); Sol Flax & Associates (electrical); Chen & Associates (soils).* Landscape: *Alan Rollinger.* General contractor, construction manager: *Pinkard Construction Co.*

The mini-megastructure that is the North Campus of Denver's Community College flaunts its solar collectors with an architectural aplomb rarely manifested by mechanical systems.

Early planning meetings for new quarters to replace the college's bursting temporary facilities took place in the summer of 1973, when warnings of impending shortages were given substance by long lines of motorists haunting still-pumping gas stations.

Thus alerted, architect John Anderson and the client commissioned the Albuquerque mechanical engineering firm of Bridgers and Paxton, pioneers in work with solar energy, to prepare a feasibility study including a solar option. The choice—later rendered moot when the local utility served notice of a cutoff of new natural gas allocations—narrowed quickly to conventional gas-heating and cooling versus solar-assisted heat pump system. The solar system commanded an

initial premium at 8.5 per cent over the original budget but was deemed a viable long-term option because of the relatively short payback period— 11 years on the conservative assumption of a 300 per cent increase in natural gas prices between 1973 and 1990.

The system as installed (comfortably within the revised budget) is, as Anderson notes, "nothing trailblazing." Playing safe with a fledgling technology, it employs 35,000 sq ft of steel flat-plate liquid collectors and water storage. Two centrifugal chillers needed for cooling serve as heat pumps to boost stored water temperature when it falls below the 100F required for the air-handling system. Back-up is provided by the domestic hot water boiler. In a typical insolation year, the system is expected to supply some 80 per cent of heating requirements.

The designers expect in addition a substantial bonus through use of the heat pumps to redistribute ambient heat gener-

Slope and framing of collector array supported by monitor structure is continued at main entrance by window wall which admits daylight to lobby and adjacent corridor-cum-"street" and, via a balconied light well, to a student lounge area below. Structure, exposed throughout, combines cast-in-place and precast concrete and concrete block. Precast double-tee roof framing members, supported elsewhere by beams, are carried at the monitor by massive concrete panels with sloping ends slotted to receive them.

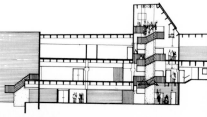

ated by people, lights and machinery, buttressed by a full gamut of heat-recovery devices. The resulting heavy heat traffic flow is directed by sophisticated control and monitoring instrumentation that Anderson and Bridgers agree is the *sine qua non* of successful system operation.

Anderson further stresses the thermal "buttoning-up" of the building as the essential point of departure for solar heating.

Particularly stringent constraints were imposed by a curriculum weighted heavily to occupational education.

On the south, flanking the main entrance, are single-story units housing shops whose requirements for loading access and heavy equipment loads mandated their placement on grade. On the north a parallel element, two stories stair-stepping to three, contains additional academic spaces, with administrative, service and support areas.

Linking the two is a 28-ft-wide central spine that serves as a continuous interior street running the length of the building, and doubles as a buffer isolating the noise and vibration of the shops from areas opposite. Its elevated roof, angled at 53 degrees on the north side, provides surface support for the north bank of solar collectors and encloses eight fan rooms containing the bulk of the building's air handling equipment. The 10-in.-thick precast panels that span the monitor section at each bay transmit lateral loads across discontinuous roof diaphragms to concrete block shear walls. Where not occupied by fan rooms, the elevated monitor provides clerestory light and dramatic vertical space to the wide "places in the road" that punctuate the structure's length and serve as student gathering places.

But it is the use of the "unused" areas of the highly—and multi—functional monitor to create the building's soaring, climactic spaces that perhaps best exemplifies Anderson's efforts to add spice without resort to prettifying or pretense.

Robert E. Fischer photos

Portholes on north wall of monitor supplement lighting from industrial-style fixtures in cafeteria area—the campus's Great Hall. Repeated in transverse panels, portholes minimize their bulk visually as well as physically.

Freestanding collectors are supported by steel frames every 9 ft and by a space-frame section (above and left) spanning 84 ft. Frames are anchored to double tees with flanges beefed up to handle wind loads.

Skylit, barn-red "silos" at building ends form playful enclosures for access stairs to roof between monitor and south bank of collectors. Indented circle on end facade of monitor echoes interior porthole motif.

Sun control
in an office building:
a passive solar system

FAMOLARE HEADQUARTERS, Brattleboro, Vermont. Owner: *Famolare Inc.* Architects: *Banwell, White & Arnold.* Consulting engineers: *R.D. Kimball (mechanical/electrical); Carroll Lawes (structural).* Solar consultant: *A.O. Converse.* General contractor: *O'Bryan Construction Co., Inc.*

The positive effect of pleasant work space on worker productivity and morale has not been lost on Joe Famolare, the exacting president of Famolare Shoes Inc. When Famolare decided to move operations from leased warehouse space in New Jersey to the mountains of Vermont, he requested attractive work space with a sharp eye kept on energy consumption.

After preliminary designs for an elaborate high-technology solar collector system were declared not cost-effective, the architects created a passive solar system for the new headquarters, using sun scoops, skylights, and special window shutters to take

Robert Perron photos

maximum advantage of natural light and sun.

Set amid farms in mountainous Vermont, the office building is topped by a large sun scoop yawning open to the south. Manually-adjusted fiberglass panels bounce sun into the scoop in winter, flooding the offices with light, and are turned to deflect sun in summer. Several smaller skylights brighten showroom areas of the office building and movable fiberglass panels hung inside the skylights can be used to deflect sun.

A system of monitors on the adjoining warehouse roof admits sun in winter for direct solar gain, and excludes sun in summer by use of overhangs. The monitors provide sufficient warehouse lighting about 80 per cent of the time; high-pressure sodium fixtures are used only in recessed areas or on overcast winter days.

Office building windows on the south are fitted inside with special sliding shutters of translucent fiberglass. The shutters are closed when the building is unoccupied to insulate the glass, and are used to exclude direct sun from offices facing south. When the shutters are closed, solar-heated air in the space between the shutter and window is collected and transferred in winter to the north side of the building. The window shutter system is estimated, conservatively, to reduce the building's heating and air conditioning load by 10-15 per cent annually.

To encourage communication between the office and warehouse personnel, even in harsh Vermont winters, a suspended tunnel of bronze-tinted acrylic plastic is used to connect the buildings. Heated air in the tunnel is collected for circulation throughout the rest of the building.

Energy conserving measures extend to the use of 6-inch thick fiberglass insulation in the office building walls, and an insulation of 2.35-inch thick isocyanurate foam board in the roof. Task-ambient lighting in the offices is designed to use only 2.2 Watts per sq ft.

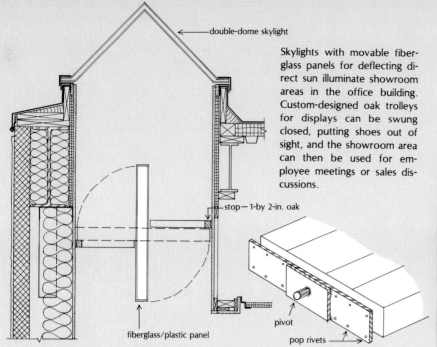

double-dome skylight

stop—1-by 2-in. oak

fiberglass/plastic panel

pivot

pop rivets

Skylights with movable fiberglass panels for deflecting direct sun illuminate showroom areas in the office building. Custom-designed oak trolleys for displays can be swung closed, putting shoes out of sight, and the showroom area can then be used for employee meetings or sales discussions.

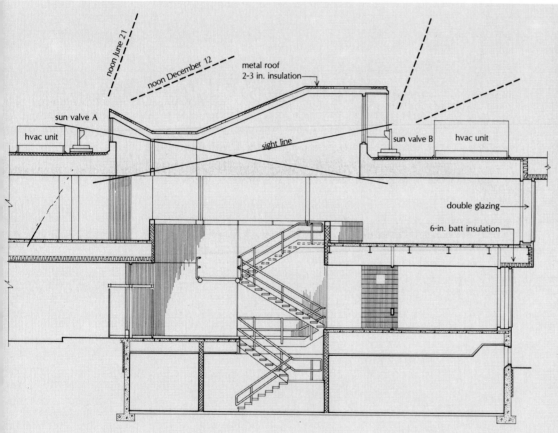

noon June 21

noon December 12

metal roof
2-3 in. insulation

sun valve A

hvac unit

sight line

sun valve B

hvac unit

double glazing

6-in. batt insulation

Movable rooftop panels (called "sun valves" by solar consultant Professor A.O. Converse of Dartmouth) tip sun through a large scoop into the office building, creating a bright center well and airy open office space (photo below far right). Hvac units on the roof near the scoop are placed so as to be out of sight from inside the building.

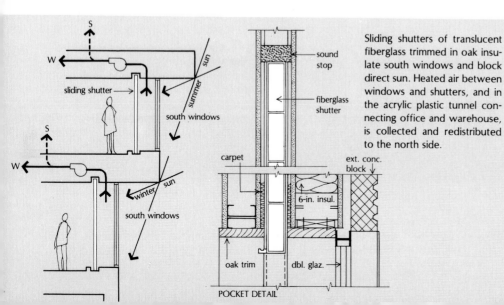

S
W
sliding shutter

south windows

S
W

winter sun

south windows

sound stop

fiberglass shutter

carpet

ext. conc. block

6-in. insul.

oak trim   dbl. glaz.

POCKET DETAIL

Sliding shutters of translucent fiberglass trimmed in oak insulate south windows and block direct sun. Heated air between windows and shutters, and in the acrylic plastic tunnel connecting office and warehouse, is collected and redistributed to the north side.

Robert Perron photos

Solar for cooling: an active solar system

MEMORIAL STATION, Houston, Texas. Owner: *U.S. Postal Service—Bill Wright, Gordon Steger, John Wright, liaison.* Architects: *Clovis Heimsath Associates—Ray Wobbe, project architect; Harold Carlson, supervising architect.* Engineers: Nat Krahl & Associates (structural); Timmerman Engineers, Inc. (mechanical). *Contractor: Westpark Construction Company.*

Because this postal facility is in a large shopping center visible from the Southwest Freeway in Houston, thought of by many as the energy capital of the country, and because he was encouraged by the Postal Service to stesss energy conservation, architect Clovis Heimsath felt that the inclusion of a solar system would be an appropriate symbol, and that patriotic colors would provide an appropriate palette.

Two major considerations dominated the design: First, only off-the-shelf technology could be considered for solar because the building had to be kept on schedule. Secondly, the solar system was possible from cost and size standpoints only through substantial reduction in load derived from energy-conservation techniques. These included: 1) reducing makeup air; 2) zoning spaces so the workroom could be substantially closed down much of the workday; 3) facing all glass north and using it only for the lockbox lobby; 4) using a square plan to reduce exterior wall area; 5) using mercury HID lamps for the workroom; 6) using insulated metal wall and roof panels to reduce heat flow.

By saving money on the building exterior and simplifying every element of design, including exposing the structure, lights and ducts, the designers could afford the additional first cost of solar, while still bringing in the building at about $40 per sq ft (less site) which was under budget (1978). Premium for the solar energy addition was $130,000 (using a central station water-cooled-condenser system for comparison). With a $6,144 per year estimated operating cost savings, the payback at current energy costs would be 21 years. Projected increases in energy costs would reduce this payback to nine years. The solar collector system is designed to provide 80 per cent of the cooling requirements.

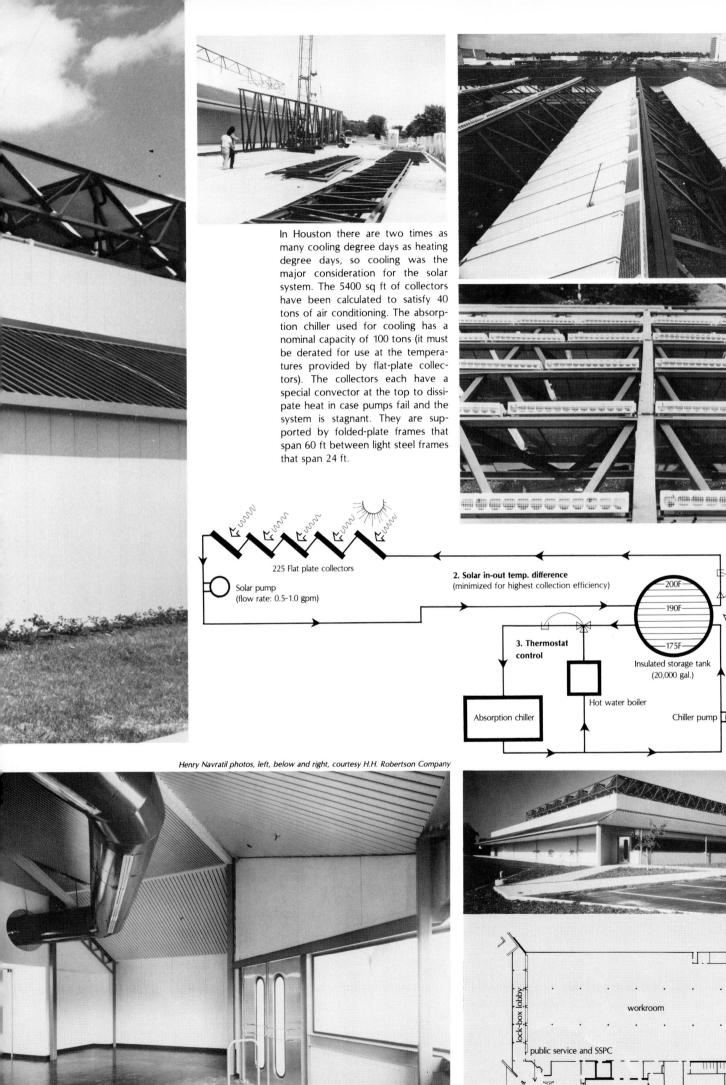

In Houston there are two times as many cooling degree days as heating degree days, so cooling was the major consideration for the solar system. The 5400 sq ft of collectors have been calculated to satisfy 40 tons of air conditioning. The absorption chiller used for cooling has a nominal capacity of 100 tons (it must be derated for use at the temperatures provided by flat-plate collectors). The collectors each have a special convector at the top to dissipate heat in case pumps fail and the system is stagnant. They are supported by folded-plate frames that span 60 ft between light steel frames that span 24 ft.

225 Flat plate collectors

Solar pump
(flow rate: 0.5-1.0 gpm)

2. Solar in-out temp. difference
(minimized for highest collection efficiency)

1. Solar control

200F
190F
175F

Insulated storage tank
(20,000 gal.)

3. Thermostat control

Hot water boiler

Absorption chiller

Chiller pump

Henry Navratil photos, left, below and right, courtesy H.H. Robertson Company

lock-box lobby

workroom

public service and SSPC

Active and passive solar systems combined

ARMED FORCES RESERVE CENTER, Norwich, Connecticut. Owner: *State of Connecticut Department of Public Works and Connecticut Military Department—Major General John F. Freund.* Architects: *Moore Grover Harper, PC—Robert L. Harper, William H. Grover, Glenn W. Arbonies, Jefferson B. Riley, Charles W. Moore.* Engineers: *Besier and Gibble (structural); Helenski Associates (mechanical).* Consultants: *Arthur D. Little Company (solar).* Contractor: *F. W. Brown Co.*

It is clear upon examining this armory in eastern Connecticut that, when architects stop to think about it, energy-conserving features can evolve as a very natural part of building design. Here Moore Grover Harper have disposed spaces naturally by function and size in a way that assists energy conservation, and that leads to logical application of both active and passive solar heating techniques. For instance, the main building steps up in three tiers, allowing logical placement of solar collectors overhanging three walls so they act as sunshades in summer. And at the far corner, the architects, in a droll step, designed a quoined tower that heralds the entrance while also serving as a chimney.

Both the State of Connecticut (which financed 25 per cent of the project and pays the operating costs) and the military wanted energy-conserving techniques, including so-

lar. The 10-acre site in Norwich Industrial Park was selected because it was clear of trees to the south, but had woods on the north to serve as a windbreak. The project comprises an Armory Building with offices, drill shed, classrooms and rifle range and a separate Organization Maintenance Shop (OMS) for vehicle maintenance and repair. Because the offices and OMS building are in continuous use, while the drill hall and ancillary spaces are used only occasionally, the solar-heating systems are functionally separate. Only the office-section solar system utilizes storage (a 2000-gal. tank). The two other systems for the armory and for the OMS building feed directly to air-handling units.

The project cost $1,459,147 (1977) of which solar heating systems cost $88,489 and other energy-saving components, $54,600. It was 25 per cent under budget.

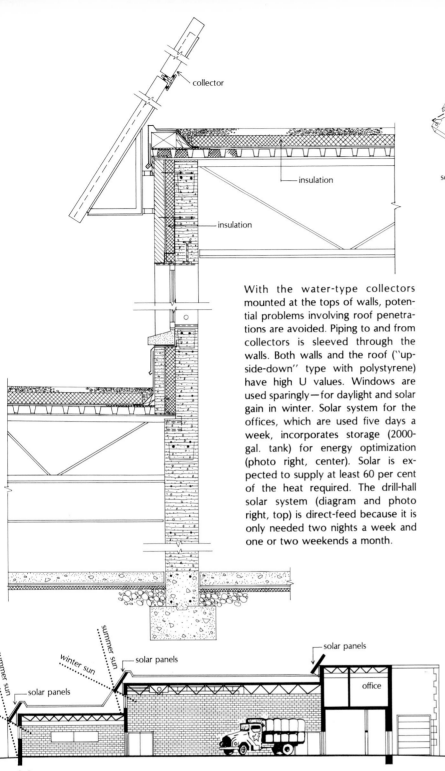

collector

insulation

insulation

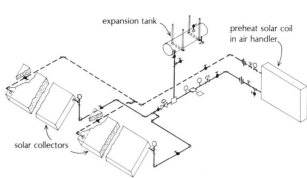

expansion tank

preheat solar coil
in air handler

solar collectors

With the water-type collectors mounted at the tops of walls, potential problems involving roof penetrations are avoided. Piping to and from collectors is sleeved through the walls. Both walls and the roof ("upside-down" type with polystyrene) have high U values. Windows are used sparingly—for daylight and solar gain in winter. Solar system for the offices, which are used five days a week, incorporates storage (2000-gal. tank) for energy optimization (photo right, center). Solar is expected to supply at least 60 per cent of the heat required. The drill-hall solar system (diagram and photo right, top) is direct-feed because it is only needed two nights a week and one or two weekends a month.

Norman McGrath photos

summer sun
summer sun
winter sun
summer sun
solar panels
solar panels
solar panels
office
solar panels

A·A

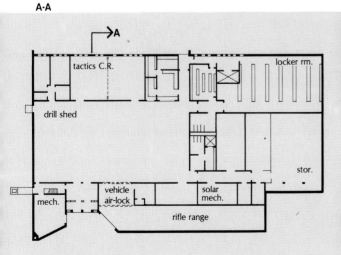

A

tactics C.R.

locker rm.

drill shed

stor.

vehicle
air-lock

solar
mech.

mech.

rifle range

Ventilation load is reduced by air locks at entrances. An overhead door at the rifle range and one at the drill shed perform this function for vehicular movement. Temperature is allowed to float in the drill shed when it is not in use: heat is added when it drops below 50F, and exhaust fans go on above 75F. The masonry walls act as thermal storage.

Improved lighting and sun control in a post office

GRANVILLE W. ELDER STATION (NORTH SHORE), Houston, Texas. Owner: *U.S. Postal Service—Bill Wright, Gordon Steger, John Wright, liaison.* Architects: *Clovis Heimsath Associates—design/production: Clovis Heimsath, Emmett White, Ray Wobbe, Jerry Mendehall, Harold Carlson.* Interiors: *Lacey Flagg.* Engineers: *Nat Krahl & Associates (structural); Timmerman Engineers, Inc. (mechanical).* Contractor: *Capellen Construction Company.*

When architects Clovis Heimsath Associates were commissioned to design a postal service facility for a rapidly developing area northeast of Houston, they were urged to make this a cost study in energy-saving opportunities. For the resulting building, the architects and their engineers predict an energy savings of close to 40 per cent plus a reduction in cooling demand of 39 per cent, when compared with a "baseline building model." They developed the baseline building from a plan type of the U.S. Postal Service and from its design guidelines. Systems specified by the Postal Service for this size building in Houston are air-cooled electric compression refrigeration and electric resistance heating.

Importantly too, the architects have achieved operational efficiencies in the layout of the facility, a more efficient lookout gallery, and much more pleasant spaces for the workers and customers.

The largest savings (16 per cent, of which 60 per cent is power for lights, and 40 per cent power for cooling) resulted from using improved-color mercury HID lighting in the workroom, and by shutting off some lights in the carriers' sorting area which are not needed for two-thirds of the time.

The next largest savings (10 per cent) came from ventilating the locker room and toilets with exhaust air from the workroom. Another 8 per cent was lopped off the baseline building by reducing the U value of the roof from 0.12 to 0.06 and the walls from 0.20 to 0.07. Savings were less for changes in fenestration, inasmuch as there was not much glass in the baseline building. Nonetheless, glass is used much more effectively here. All windows face north in sawtooth fashion along the lockbox lobby, and at the clerestory above the workroom, giving a more pleasant space plus daylight.

Ed Stewart Photography & Associates, Inc. photos

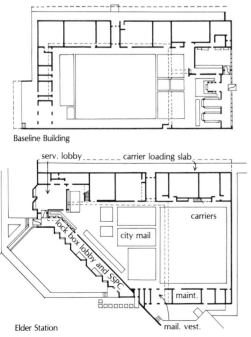

Baseline Building

serv. lobby — carrier loading slab

lock box lobby and SSPC

carriers

city mail

maint.

mail. vest.

Elder Station

Overhangs and rooftop collectors: an active solar system

ALABAMA POWER COMPANY DISTRICT OFFICE, Montevallo, Alabama. Owner: *Alabama Power Company*. Architects: *Cobbs/Adams/Benton— Doyle L. Cobb, principal; Aubrey Garrison III, project architect*. Engineers: *Miller & Weaver, Inc. (mechanical); Cater & Parks, Inc. (electrical)*. Consultant: *Vachon & Nix, Inc. (solar)*. Contractor: *Andrew & Dawson*.

In their design of a 17,000-sq-ft office building near Birmingham for the Alabama Power Company, an energy-conservation demonstration project for the utility, the architects discovered not only that saving fossil fuels was not as forbidding as they had first thought, but also that they could develop a new esthetic in the process.

From a planning standpoint, the building was skewed at the front end of its long, narrow site so there would be room for a pedestrian plaza and for visitor parking, and so that the rear of the site where company vehicles are parked would be blocked from view. The 2½-floor building has spaces for operation/marketing, accounting, display/auditorium and appliance repair.

With the building oriented so that the long wall faces directly south the rooftop solar collectors could be mounted parallel with the front. To keep sun from entering, however, was a challenge for the architects which they solved through a study of overhang and vertical-fin configurations, and with an awning for the entrance. The building was selected by ERDA (DOE) as one of the first 32 projects to be funded by the Solar Demonstration Act of 1975.

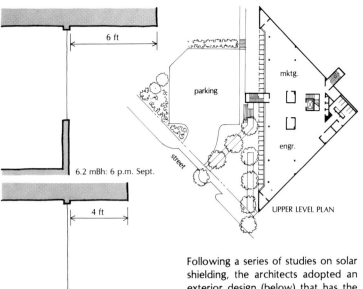

6 ft

6.2 mBh: 6 p.m. Sept.

4 ft

8.1 mBh: 12m Dec.

4-ft fin.

7.2 mBh: 12m Dec.

6-ft fin.

5.1 mBh: 4 p.m. July

UPPER LEVEL PLAN

parking

mktg.

engr.

street

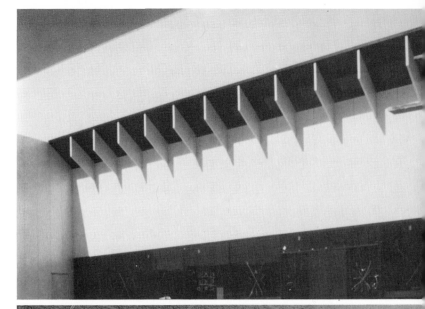

Following a series of studies on solar shielding, the architects adopted an exterior design (below) that has the glass sloped 32 degrees away from the building, fins 4-ft apart, and a 4-ft overhang. The tilt of the glass is the angle of the sun at about noon in December. The fins block the sun from southeast and southwest. The fenestration has a sill height of 3½ ft above the floor and a head height of 6 ft 4 in., so a person can see out whether seated or standing.

Daylight is supplemented by indirect fluorescent lighting at the perimeter, and desk lights plus indirect lighting are used for the interior.

The exterior wall is a 2-in. prefabricated, foam-filled metal panel with a U value of 0.08.

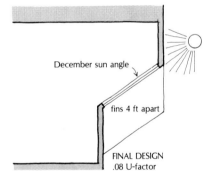

December sun angle

fins 4 ft apart

FINAL DESIGN
.08 U-factor

Henry Navratil photo, courtesy H. H. Robertson Company

The solar-assisted system comprises 2500 sq ft of flat-plate collectors using a black chrome finish, an 8000-gal. hot water storage tank, a nominal 25-ton absorption chiller, inert gas freeze-protection system, and control and data-monitoring equipment. To enhance the efficiency of the collectors for cooling, which are mounted at 30 deg., the system has 8-ft-high front mirrors set at 45 deg., and top mirrors, 2-ft high. Standby heating (100 per cent) is provided by an electric boiler. Solar cooling is available when the solar water temperature is above 170F. The remaining time, a 30-ton reciprocating water chiller is utilized which has an 8000-gal. tank, sized for offpeak storage of chilled water. Freeze protection is accomplished by draining the collectors and replacing the water with nitrogen gas. Premium for the solar was about $170,000 (1977).

Courtesy Butler Manufacturing Company

Collectors as walkway canopies: an active solar system

ARTS VILLAGE, HAMPSHIRE COLLEGE, Amherst, Massachusetts. Owner: *Trustees of Hampshire College*. Architects: *Juster Pope Associates*. Consultants: *Cosentini Associates (mechanical engineering); Joseph Frissora (energy)*. Landscape architect: *V. Michael Weinmayr*. Construction manager: *Edward J. O'Leary Co., Inc.*

High performance tubular solar collectors form graceful blue-canopied walkways linking simple but tastefully-detailed buildings of pre-engineered components in this new Arts Village at Hampshire College.

Architects for what eventually will be a five-building arts complex at the experimental college wanted the solar system to serve the entire complex, despite a phased construction schedule. At the same time, they wanted to incorporate the solar system as a major design feature of the complex in order to put the sometimes obtrusive collectors back in proportion to the buildings. Consequently, the rows of all-glass, selectively-coated evacuated tubular collectors were transformed into a string of canopies on a bright blue steel structure between buildings. The result was a functional but clean "environmental sculpture" overhead and large sheltered building walkways below.

The design solution solved several practical problems of building with the solar elements as well. Moving the collectors from building roofs eliminated extra load which could have prohibited use of the simple buildings of pre-engineered components, and allowed use of a standing-seam metal roof system.

The solar system, made possible largely by a $355,000 Federal grant, is expected to supply 95 per cent of the complex's cooling needs, 65 per cent of the space heating requirements, and all domestic hot water. Coordination of the solar system is done by a microprocessor. Additional energy not supplied by solar will be obtained from a central electric boiler in the complex.

Because of the build-as-funded nature of the project, and the need for flexible work space that could change with curriculum and teaching techniques, the architects used understated but well-detailed pre-engineered building components that are complemented with landscaping and the solar canopies.

Two of the five buildings have been completed to date, with the Painting Studio taking just 10 months from design to occupation, and the Music and Dance building requiring a similar construction timetable.

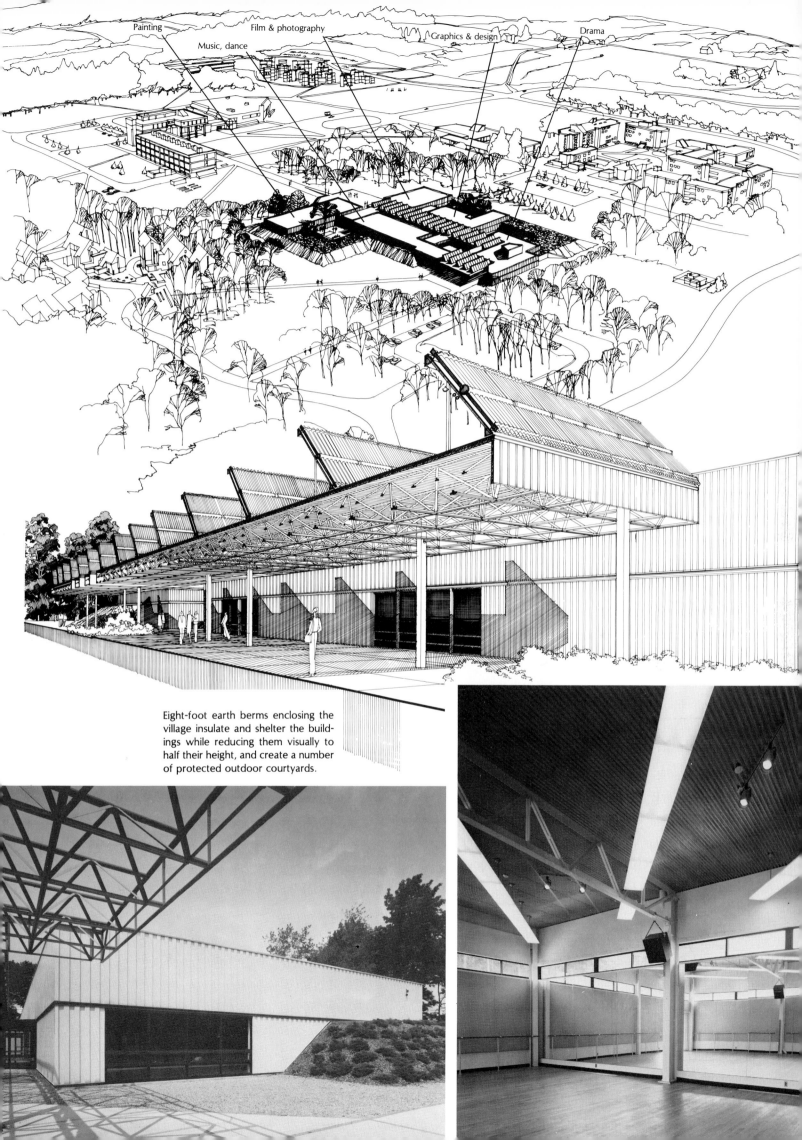

Painting

Music, dance

Film & photography

Graphics & design

Drama

Eight-foot earth berms enclosing the village insulate and shelter the buildings while reducing them visually to half their height, and create a number of protected outdoor courtyards.

CHAPTER 6:

New Approaches to Lighting and Integration of Lighting and Mechanical Services

The discipline of architectural lighting has not wanted for tools to work with, but it is only recently that a deeper understanding has developed as to how these tools might best be used to produce the right kind of light to see by ("task-oriented" lighting) and the right kind of light for the visual environment. Much interest has been generated in furniture-integrated lighting, and some significant examples of this approach are presented in this chapter. But task-oriented lighting can take a number of forms. This chapter, for example, also presents two outstanding examples of flexible ceiling lighting systems that can be changed around as task situations change; one is an office building and the other a library.

Not since the 50's have building designers shown as much interest as now in daylighting for seeing tasks, such as reading. This chapter presents an example that illustrates the aesthetic potentials of daylighting as much as its practicality.

Talented lighting designers also have recognized the potential of manipulating light via more sophisticated lighting control devices such as lenses and reflectors to suit particular building uses and to save energy; this chapter shows two outstanding examples.

And, finally, more attention is being paid by architects and engineers to a more rational integration of lighting, wiring, and mechanical services—using structure and other building design elements multifunctionally.

Task/ambient lighting

To help translate employee needs into the fabric of furniture and lighting for ARCO (Atlantic Richfield Company, Philadelphia), the interior design firm of Interspace, Incorporated involved environmental psychologist Ronald Goodrich early in their programming studies. Now, after two years' use, ARCO asked Goodrich to find out how well employees like it. Goodrich issued a questionnaire to 20 per cent of employees, top management to secretaries, and got an unusually high 70 per cent return. "The replies," Goodrich says, "showed that lighting

received the most positive response of any element of the design, save for the indoor plants." More of his analysis follows later.

As the programming studies developed, Interspace also brought in lighting consultant Sylvan R. Shemitz, who conceived the lighting approach and developed both the built-in lighting and the kiosks for uplighting and accent lighting. Guiding the project for the Atlantic Richfield Company was staff engineer Ben Cubler, who, as a committee-of-one, not only looked after client objectives and project scheduling, but also

was involved with lighting quality goals, and with such nitty-gritty as ease of setup and relocation of the furniture, detailed in metal and wood by JG Furniture.

In Goodrich's questionnaire, four questions dealt with how employees felt about the lighting . . . whether the characteristic mentioned mattered, "to some extent, to a moderate extent, or to a great extent." Question 1: Does this type of lighting create a more pleasant environment? Question 2: Is [a room with] this type of lighting easier to work in than one with

NO FIXTURES   FIXTURE AT DESK   UPLIGHT FROM WORK STATION   UPLIGHT FROM CABINET

standard ceiling lighting? Question 3: Does this lighting create less visual fatigue than standard ceiling lighting? Question 4: Does this lighting create a visually more attractive environment?

While Goodrich has not analyzed all the data, nor examined all the variables, his tentative conclusions from replies to the questionnaire are: Employees generally feel that the task lighting creates a more pleasant environment and makes it easier to work. Analysts and secretaries in interior spaces, some of whom are blocked off from windows,

gave lower ratings to the lighting as "creating less visual fatigue." Administrative employees on the perimeter were most positive on this question, while management personnel with private offices were less positive (75 per cent said it was true "to some extent;" 40 per cent said it was true "to a great extent"). Consideration of these replies, says Goodrich, suggests that: a) windows influence people's judgment about the standard of lighting, and b) different levels of employees have different conclusions about lighting—i.e., how much difference is perceived. In any event,

Goodrich says that more investigation of this question is necessary.

There were indications that younger people feel more positive about their environment in general, and about the lighting in particular. And people who liked their jobs, and described their jobs as non-routine, rated the visual environment higher.

Of all the design elements used in the offices, preference data showed that the task and ambient lighting was ranked either one or two as being the most important.

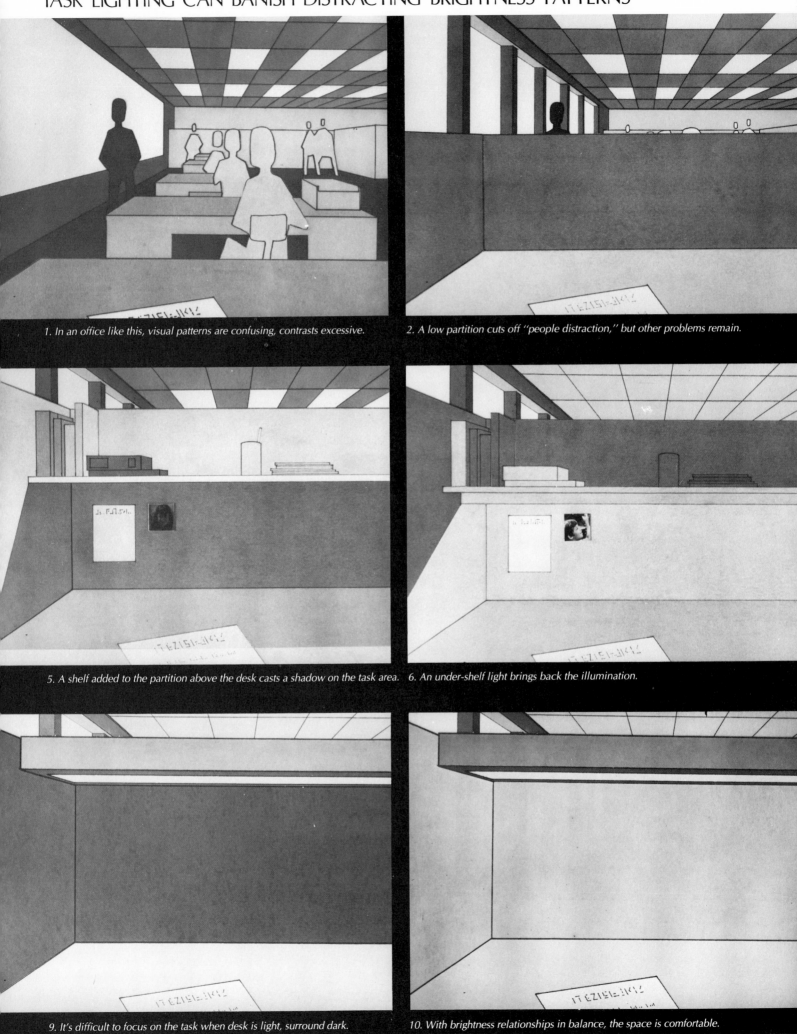

1. In an office like this, visual patterns are confusing, contrasts excessive.

2. A low partition cuts off "people distraction," but other problems remain.

5. A shelf added to the partition above the desk casts a shadow on the task area.

6. An under-shelf light brings back the illumination.

9. It's difficult to focus on the task when desk is light, surround dark.

10. With brightness relationships in balance, the space is comfortable.

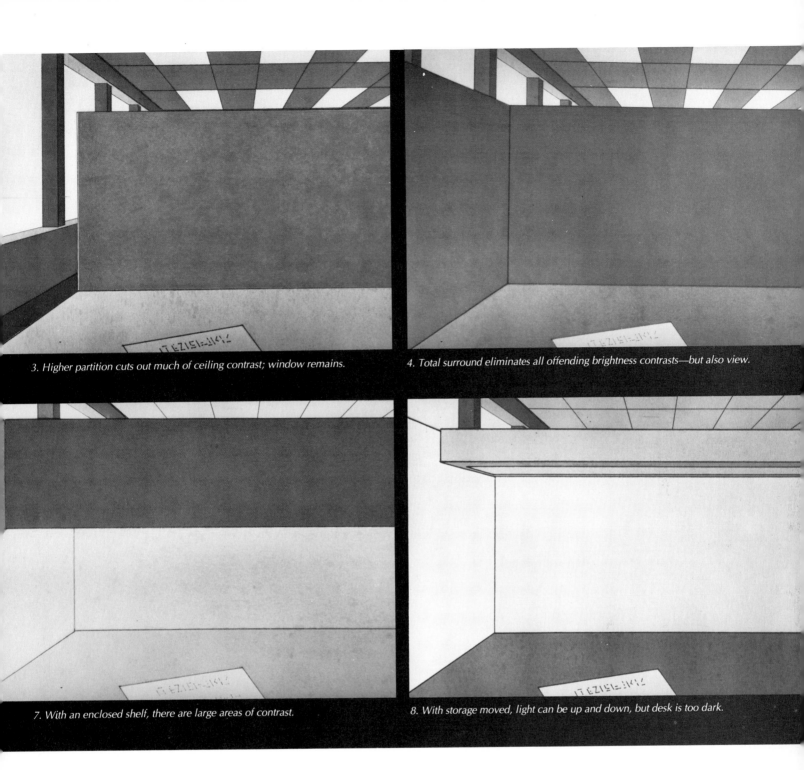

3. Higher partition cuts out much of ceiling contrast; window remains.

4. Total surround eliminates all offending brightness contrasts—but also view.

7. With an enclosed shelf, there are large areas of contrast.

8. With storage moved, light can be up and down, but desk is too dark.

Lighting consultant Sylvan R. Shemitz uses these black-and-white renderings to convey some of the visual-comfort problems common to many conventionally lighted offices, to indicate some the pitfalls on the way to getting rid of visual distractions and annoying ceiling brightness contrasts, and to emphasize the benefit to be gained from balanced brightness relationships via the task/ambient lighting approach.

In office design, it always has been good practice from the standpoints of lighting efficiency and visual comfort to use light colors for room finishes and for furniture, except where the tasks are not very critical. But, as Shemitz points out, this does not imply a bland, lifeless interior environment. It does mean, however, careful attention to color schemes and textures, particularly with respect to the work stations and ceiling surfaces. While color of the carpet or the floor finish is not as critical from a contrast standpoint—though a marked difference could be distracting—it does have a significant effect upon the efficiency of the ambient lighting. The darker the floor, the less ambient light there will be overall. For example, in a task/ambient lighting installation for Texas Eastern Transmission in Houston, a light beige linen fabric was used for the vertical surface back of the uplight/downlight, and the work surface and the rest of the surround are light oak.

Once room distractions have been disposed of, the lighting design problem focuses on brightness relationships within work station furniture. The goal, says Shemitz, is to avoid striking brightness contrasts between work-station surfaces so that the eye does not need to make drastic adjustments when shifting away from the task and back to it. This involves control of illumination and reflectances.

When ambient lighting is provided by uplights within the furniture and from freestanding units, the lighting must be carefully designed to avoid "hot spots" of light on the ceiling which could be just as disconcerting as any checkerboard of bright ceiling luminaires and dark ceiling surround.

Jeremiah O. Bragstad

Lighting for space definition and visual comfort

Employees and visitors alike must yield to the charms of the interior and exterior vistas that architect Matthew Mills has created for the corporate offices of Crowley Maritime Corporation in San Francisco. Deft planning and deft use of glass open up views to the outdoors for everyone, even from rooms in the core, while the interior spaces—designed by the architect with a restrained palette of materials, surfaces that are light or dark or reflective, and flush detailing of all design elements (even emergency lights)—make the offices pleasant and interesting and easy to work in. The plan's calculated simplicity is underscored by the reserved use of materials—white oak, glass, and light carpet. The lighting has been inventively disposed for space definition, visual comfort and work activity (at low energy consumption) by consultant Sylvan Shemitz and architect Mills—a truly successful collaboration of an experienced consultant and an architect who knew exactly what he wanted. Throughout the day the glass is alive with reflections and transparencies that constantly change. It is a delight, says Mills, to look through a reflection of the Bay Bridge into the glow of an adjoining space. "For me it is these many layers of subtlety that are exciting because they allow the mind to move effortlessly from one to the other as the mood warrants, without ever losing an over-all sense of place."

Visitors enter the main reception area on the executive floor though frameless glass doors.
Employees leaving the elevators may turn either right or left to reach their offices. Fluorescent
wall-washers illuminate the elevator lobby, reception area and core walls. Staff groups
occupy three sides of the plan, and only two exterior walls were taken for the 16 executive offices.

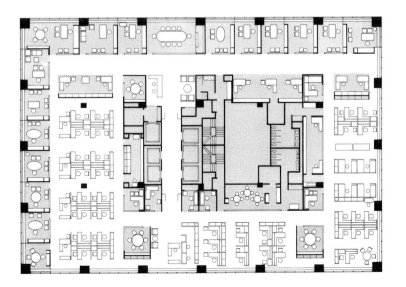

The corporate offices for Crowley Maritime Corporation are located in One Market Plaza, a high-rise office complex at the foot of Market Street, adjacent to the historic Ferry Building and overlooking all of San Francisco Bay. Crowley operates the world's largest fleet of tugs, barges and offshore exploration vessels, and has always had its offices on the waterfront. In a move to consolidate its corporate operations and accommodate recent expansion, Crowley chose its new location because of its competitive rent and its unrestricted views of the bay and harbor. Occupancy will total about 45,000 sq ft on the 38th (executive), 33rd and 32nd floors; plans are also under way for the 31st floor. With the commitment to move came the opportunity to improve planning. Architects Robinson and Mills and the Crowley staff worked closely to find concepts that would yield a flexible plan responsive to ever-changing work relationships.

In late 1975, the architects proposed that Crowley utilize a reasonable mix of open planning and private spaces, and an improved

185

From the reception area, the overlapping of interior reflections and the nighttime panorama of San Francisco harbor create an enigmatic multiple image in the flush-glazed partition of the board room and in the outside window. The interior glass wall, which also encloses executive offices, is terminated by an oak-surfaced header at its top and by an oak sill at the bottom.

generation of the task lighting system. According to Crowley's project coordinator, Lee Ready, the company sought better work flow, extended life, and a generally improved environment—"a space where the office environment can continually evolve." Crowley's decision to accept the proposal was based on the value of improved lighting and related energy savings (only 1.8 watts per sq ft connected load), and on the flexibility of a modular lighting system. The dual task for the architects was to develop a plan that would accommodate crucial functional relationships, and at the same time reflect the Crowley organization in a quiet sense of simplicity and quality.

The executive floor (77 people), unlike many company headquarters, was to be a "working" floor not only for the senior officers, but also for administrative, engineering and support staffs—with all working spaces organized in arrangements that would encourage interaction. To achieve this, the architects developed a space with an expansive and transparent quality that gives every employee natural light and a view. The plan (right) provides a variety of open and private spaces around an expanded core containing spaces for common use, and joined by a simple and efficient circulation system.

The plan comprises three basic space types: 1) an expanded core, 2) private perimeter offices, and 3) open-plan area. In the center of the core is the elevator lobby (seen directly and reflected, above). It is treated as an extension of the circulation corridor: its floor is the same light natural wool carpet used throughout the offices, and walls are detailed as in the rest of the core. The doors, required by code and for security, are frameless tempered glass. This lobby provides an added through corridor and offers the visitor an immediate view to the outside. It opens to the personnel department and to the main reception area, which is treated as a widened area in the corridor that tempts the visitor to pause and admire the panorama of San Francisco harbor on the far side of the conference room.

Illuminated surfaces provide ambient light and visual comfort, and give space definition and orientation. In private offices, a concealed uplight washes the ceilings evenly. All the core walls are washed by recessed fluorescent fixtures (lower left photo), while the glass walls of conference rooms (same photo) and the header of the executive secretary alcove (lower right) are not.

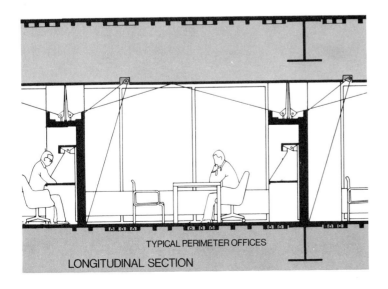

TYPICAL PERIMETER OFFICES

LONGITUDINAL SECTION

In addition to several private offices, the core includes service areas, conference rooms, a library and a coffee room. While internally complex, the core is an essentially simple form that acts as a reference plane for occupants of the surrounding space. Its walls are penetrated by doorways and strip windows, which align at the bottom with the upper surface of the open-plan work units and at the top with the headers that receive the glass partitions of executive offices and conference rooms. These headers also define secretarial alcoves, where they help contain typing noise and form a kind of nonexistent wall that discourages unnecessary visiting.

The 16 executive offices, with their continuous flush-glazed walls, partially wrap, and serve as a foil to, the white central core. In between are the open staff areas. The staff and executive offices are thus separate, yet not separate. The perimeter offices are partitioned by custom-designed "furniture-like" work units that stop short of both walls and ceiling. Privacy closures are made of acoustical glass so that

Work stations in the open-plan areas use Knoll-Stephens units that provide both task lighting and broad general uplight. A distributed file system separates the staff station from the corridor. One lighted core wall is shown in the left background. An opaque conference room wall (right background) displays photographs of the company's operations, illuminated by incandescent wall-washers.

ceiling and wall surfaces of adjacent spaces can be seen by any office occupant. Further, the glass closure to the perimeter wall permits each office to borrow its neighbor's view, greatly increasing the apparent space. Interior drapes and blinds were dispensed with in favor of the openness espoused by the company president.

Ceilings are washed by concealed fluorescent uplights above work units that provide ambient light when it is dark. The light beam from these uplights is carefully controlled so that the lighting fixture is not mirrored in the acoustical glass above the work unit. This lighting is controlled by a photocell, ensuring energy savings when daylight is sufficient.

The wall opposite each office's work unit (which is the back of the next office's work unit) has a door that sinks flush with the wall when opened and a large oak panel that can be used as a graphic display surface—all washed by recessed incandescent fixtures. Each office thus is defined by illuminated planes.

Joshua Freiwald

Jeremiah O. Bragstad

Jeremiah O. Bragstad

Joshua Freiwald

Jeremiah O. Bragstad photos

Much of the sense of quality and interest derives from the architect's and lighting consultant's attention to detail. Just a soupçon of polished brass hardware adds sparkle to the quiet elegance of the elevator lobby. And tiny brass fasteners secure the glass at sills of executive offices.

The recessed wall-washers around the core terminate at the juncture of wall and ceiling in a visor of bright aluminum (not needed for light control) that mirrors ever-changing images. Emergency lights are set flush with the continuous header of the core for appearance, and also to light corridor surfaces (top, right).

By stopping executive office work units short of the exterior wall, and providing acoustical separation with glass, the architect gave occupants ex-

panded views of the city. The photo just above looks at the outside of a corner office, showing the back of a work station. Transoms between executive offices are acoustical glass (see photo center, right). Office doors (one shown ajar in the photo) recess into door pockets so the entire oak wall can be flush.

Electrical energy is saved not just by low installed wattage: at 6 P.M. motor-driven master switches turn out all the lights except those in the core, which remain on for 15 minutes longer. This allows a late-working occupant time enough to go to a specially designed lighting panel in the core (right) to flip a low-voltage switch and turn lights back on in his area (identified by trans-illuminated numbered areas on the panel).

CROWLEY MARITIME CORPORATION HEADQUARTERS, San Francisco. Architects: *Robinson and Mills Architecture and Planning—partner-in-charge: Matthew R. Mills; project architect: Thomas Goodwin; designers: Edward Fernandez, Rae Hagner, Steve O'Brien, Richard Tobias.* Consultants: *Sylvan R. Shemitz and Associates, Inc.* (lighting design); *Jaros, Baum & Bolles* (mechanical); *Sylvan R. Shemitz and Associates, Inc.,* in association with *Marion, Cerbatos & Tomasi, Inc.* (electrical); *Wilson, Ihrig and Associates* (acoustical); *Reis and Company, and Michael Manwaring* (graphics). Contractors: *Jacks and Irvine General Contractors, Inc.* (general); *Rosendin Electric, Inc.; Architectural Wood Products.*

Lighting
and air distribution
integrated
with concrete structure

The new head office for the Bank of Canada is a multifaceted gem of green-tinted reflective glass trimmed with copper, perched on a rock escarpment across from the Canadian Parliament. By drawing on modern technology and by making use of skilled artisans (Japanese workers pre-patinated the copper trim), the building offers both a foil and a complement for nearby government buildings.

The building comprises two 12-story office towers joined by a Garden Link that also embraces one-third of the existing five-story granite bank, which will house one of the world's outstanding numismatic collections. The Garden Link has been designed for building circulation — a foyer for all three buildings, plus bridges connecting the towers — and for public amenity (for one thing, a respite from Ottawa's cold winter weather).

The skylighted area has two layers of glass between which conditioned air is circulated to prevent condensation and to melt snow in winter. Perforated copper-trimmed spandrels supply a curtain of air to the glass walls of the Garden Link. The inner portions of the engaged columns serve as vertical plenums for the spandrels. The building is sheathed with reflective double-glazing.

Because Ottawa is in Seismic Risk Zone 2 (VII on Richter scale) the Garden Link structure — essentially a stiff diaphragm on "sticks" — has its roof structure pinned to one of the office towers and has a sliding joint at the other, so it can move independently of the stiff concrete office frames.

Inside the towers, a structure of concrete "trees" on a 30-ft module incorporates the lighting, mechanical and electrical systems. The structure embodies a great number of functional and planning advantages. It provides "natural" channels for routing services, and "umbrellas" to reflect indirect light. It reduces the number of columns to a minimum, and eliminates them entirely around the windows. It provides solid concrete members into which partitions can frame. And it gives a sense of territory to small groups in the open-office scheme.

BANK OF CANADA, Ottawa. Architects: *Marani, Rounthwaite & Dick* and *Arthur Erickson*, associated architects — *Ronald Dick* (administrative partner-in-charge), *Arthur Erickson* (design partner-in-charge), *James Strasman* (project design architect). Engineers: *C.D. Carruthers & Wallace* (structural); *ARDEC Consulting Engineers* and *Brais, Frigon* (mechanical/electrical). Consultants: *William M. C. Lam* (lighting); *Emil van der Meullen* (landscape architect).

Two 12-story office towers are joined by a climate-controlled Garden Link that serves as a foyer, a bridge between the towers and as a public space.

Robert E. Fischer photos

191

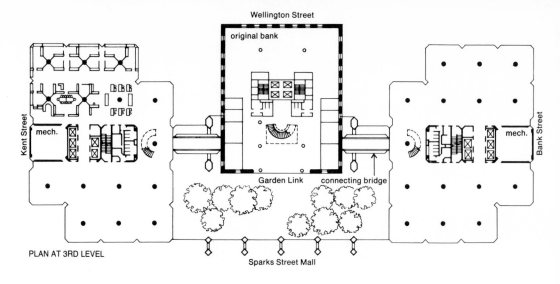

Wellington Street

original bank

Kent Street

mech.

PLAN AT 3RD LEVEL

Garden Link   connecting bridge

Bank Street

mech.

Sparks Street Mall

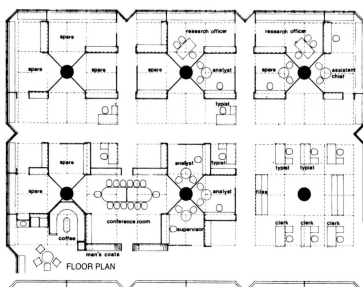

FLOOR PLAN

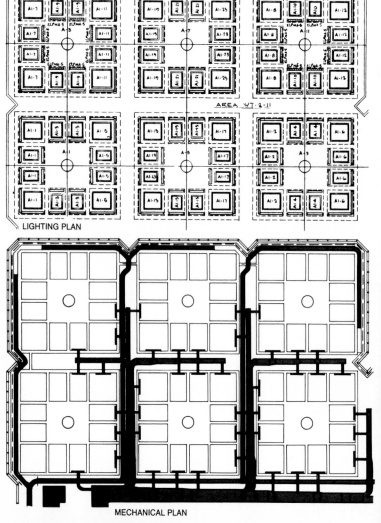

AREA WT-2-11

LIGHTING PLAN

MECHANICAL PLAN

The structural system is a series of independent structural trees, 30 ft on center. Each tree has a 3-ft-dia. column supporting a 25-ft capital composed of 3-ft-deep ribs cantilevering from a drop head in the center.

This system creates a 5-ft-wide space around each tree, providing service space for air ducts, wiring and lighting ballasts. The underside of the channel has a suspended ceiling for access. The "umbrella" portion of the trees was made 3-ft deep to accommodate services, and not for structural reasons—though this depth did save reinforcing steel. Similarly, the rib spacing was chosen not for structural reasons but to provide convenient locations for partitions and properly sized coffers for lighting.

Each cell created by the ceiling had to provide air supply and return (via slotted boxes attached to the lighting units), lighting (via direct/indirect luminaires), plug-in electrical distribution (via an electrified track around the "lid" of each coffer which allows use of accent lighting and the installation of power poles, if required), and sound absorption (via acoustical panels in the coffers). Lighting level is 90-120 footcandles (initial); office furniture is light oak, and carpet is tan/beige.

In a practice being used increasingly in Canada (for flexibility and smoke control), there are independent "package" mechanical rooms for each floor of the office towers. The low-pressure fans use only about one-half as much energy as high-pressure fans with centralized mechanical rooms.

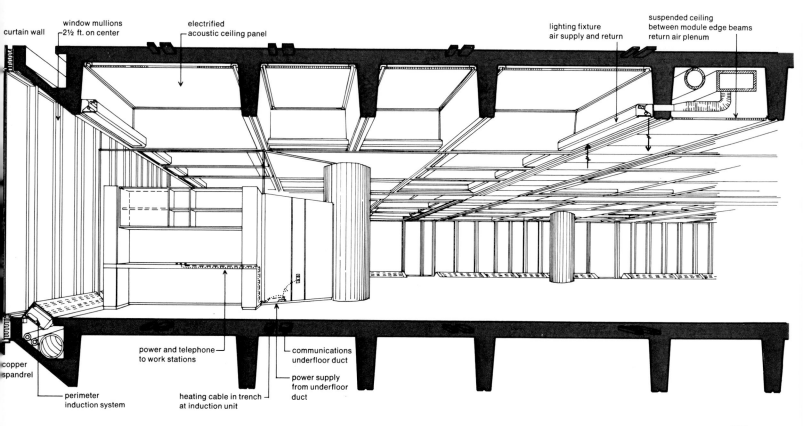

curtain wall

window mullions
—2½ ft. on center

electrified
—acoustic ceiling panel

lighting fixture
air supply and return

suspended ceiling
between module edge beams
return air plenum

copper
spandrel

perimeter
induction system

power and telephone—
to work stations

heating cable in trench—
at induction unit

communications
underfloor duct

power supply
from underfloor
duct

Task lighting and thermal storage for a large office complex

A whole series of energy-reducing steps in design, including two strikingly new departures—one in ceiling lighting, another in thermal storage—held consumption in this Canadian federal office building in Toronto to 66,000 Btu/sq ft/yr without sacrificing the quality of working space.

Indeed, few buildings whose office floors cost only $33/sq ft (1976), as did these, offer such amenities as provided by architect Macy Dubois. The zigzagged facade of this building, for example, provides more perimeter wall than a square building twice its height, and thus opens pleasant vistas for office workers. In addition, three atriums offer interior views. The main atrium, shown across page and in plan, is five stories high, and is entered from grade. Reason for the opaque wall was anticipated construction of a parking garage adjacent. Columns support the sloping roof and skylight structure, and were braced for lateral stability. Another large atrium (see top photo) will have a louvered exterior wall to exclude direct sunlight.

Undoubtedly the most dramatic advance in the design of the building is the ceiling task lighting, which requires only 2W/sq ft to provide 100 footcandles (required by union contract) on desktops. This feat is accomplished by movable, plug-in, two-lamp luminaires that can be dropped randomly into an exposed 5- by 5-ft raceway grid.

The exposed 2½- by 6-in. raceway is multifunctional: 1) it carries 347-v power (Canada) for the plug-in lighting, 120-v power for convenience outlets, and telephone wires; 2) it supports luminaires and acoustical panels; 3) it accepts linear diffusers, designed to issue up to 265 cfm per 5-ft length; 4) it accepts sprinkler heads, partition anchorages and service poles.

A large percentage of the energy savings in the 825,000 sq ft building is attributable to the design of the hvac system, which allows very large reductions in fan energy as compared with conventional installations. Moreover, a considerable amount of heat is recovered from lights, people and machines, and stored in four 75,000-gallon concrete tanks in the basement.

Reductions in fan energy are effected by 1) variable air volume distribution, 2) the use of low-temperature salvage heat in under-window radiation heaters, rather than air at the perimeter, and 3) the substitution of small air-handling units (two per floor, 26 altogether) for larger stations, substantially lessening total friction loss.

According to mechanical engineer Robert Tamblyn, the storage of salvage heat yields energy savings comparable in magnitude to the air side of the hvac system. Indications are that thermal storage will reduce heating costs by as much as 60 per cent. On the cooling side, furthermore, thermal storage makes it possible to reduce the size of the chillers by 30 per cent, and to save 20 per cent on cooling costs by cutting electric demand 30 per cent by running chillers during off-peak periods. Chilled water will be stored year-round.

GOVERNMENT OF CANADA BUILDING, Toronto. Architects: *DuBois-Strong-Bindhardt and Shore Tilbe Henschel Irwin, associated architects and planners.* Engineers: *R. Halsall & Associates Ltd.* (structural); *Engineering Interface Limited* (mechanical); *Jack Chisvin & Associates Ltd.* (electrical). Project manager: *Public Works Canada, Ontario Region.* Construction management: *E.G.M. Cape & Co. Ltd.*

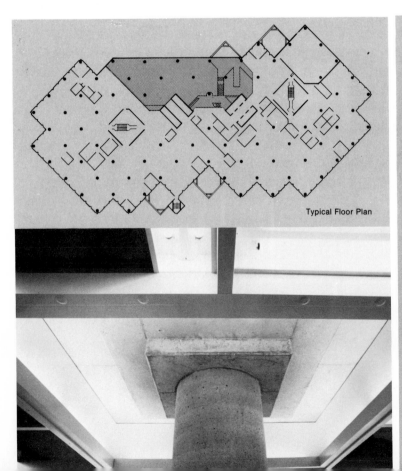

Typical Floor Plan

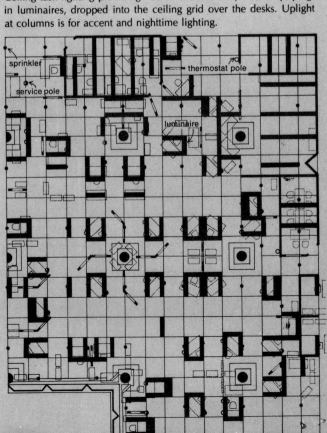

Ceiling task lighting providing 100 fc at desks is by two-lamp, plug-in luminaires, dropped into the ceiling grid over the desks. Uplight at columns is for accent and nighttime lighting.

sprinkler
thermostat pole
service pole
luminaire

Robert E. Fischer photos

GOVERNMENT OF CANADA BUILDING

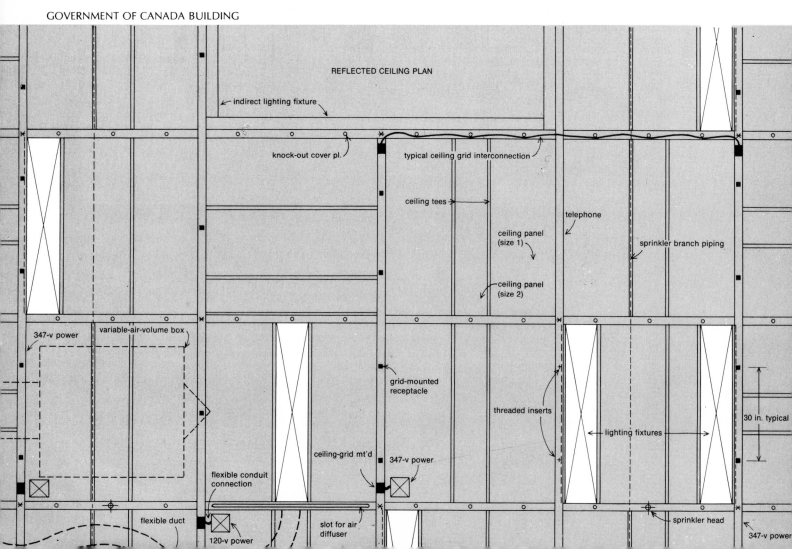

REFLECTED CEILING PLAN

indirect lighting fixture

knock-out cover pl.

typical ceiling grid interconnection

ceiling tees

telephone

ceiling panel (size 1)

sprinkler branch piping

ceiling panel (size 2)

347-v power

variable-air-volume box

grid-mounted receptacle

threaded inserts

30 in. typical

lighting fixtures

flexible conduit connection

ceiling-grid mt'd

347-v power

flexible duct

120-v power

slot for air diffuser

sprinkler head

347-v power

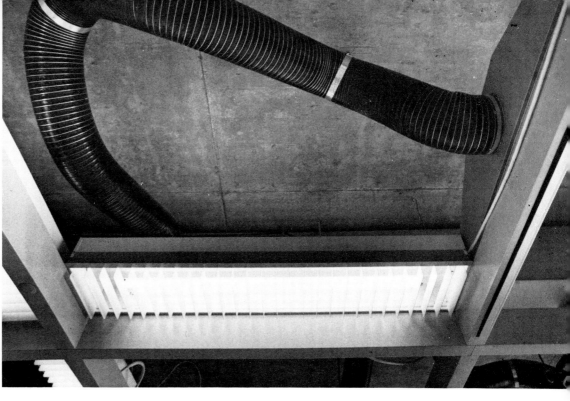

The ceiling grid performs multiple architectural and engineering functions

The first purpose of the grid system is to support luminaires for task lighting—and to allow variation in their placement according to the location of work stations. The grid further incorporates raceways to power the luminaires, which are plugged into twist-lock receptacles via power cords. Though luminaires are arranged in a more or less random pattern, the regular spacing of the 6-in. deep grid helps organize the ceiling visually. Orange-colored acoustical panels, matching the grid, come in two standard sizes to fill the 5-ft module whether one, two or no luminaires are inserted.

Every other raceway carries 347-v wiring for lights; they alternate with raceways for 120-v receptacle power and raceways for telephone wires. The 347-v raceways are on 10-ft centers because power cords for luminaires are limited by code to 6-ft lengths. Cords for 120-v power to service poles can be 12 ft, however. Telephone wires and 120-v cords to service poles are enclosed in the dummy grid members running perpendicular to the raceways. The raceway spacings thus give complete flexibility in the location of work stations.

Grid members running perpendicular to the raceways provide for a number of functions: slots for linear air diffusers, button-type knockouts for the installation of spring-loaded service poles, and round holes for sprinkler heads. The raceway grid has threaded inserts for partitions.

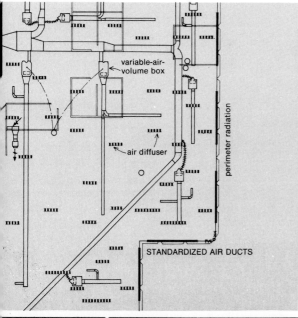

variable-air-volume box

air diffuser

perimeter radiation

STANDARDIZED AIR DUCTS

The hvac system combines energy savings with a low first-cost

Four 75,000-gallon concrete tanks like the one at left will store "waste" heat recouped from lights, equipment and people. During cold seasons, this stored heat will provide 100F water for finned radiation at the perimeter. Chillers will run in offpeak periods to reduce demand charges, and cooled water will be stored in the tanks. A problem in storing chilled water is that supply and return water may not mix, or else the supply water will not be cool enough for dehumidification. Engineer Robert Tamblyn's ingenious solution to this dilemma was to provide a plastic baffle in the shape of a huge square bag, which moves back and forth as relative volumes of supply and return water vary at night and during the day. The storage tanks cost $175,000 (1977), but saved $75,000 in chiller costs and promise to save $10,000 a year in demand charges. Six standard sections of ducts rather than many different sizes cut hvac costs. The total system bid cost under $3.40/sq ft (1977).

Robert E. Fischer photos

Three air systems and task lighting for energy conservation

In its design of the facade of the new Federal Home Loan Bank Board Building, Max Urbahn's office has been properly deferential toward the Executive Office Building across the street, which a Washington architecture critic has said is "treasured for its robust immodesty." In the facade design the architect also has been mindful of GSA's energy-use guidelines,

inasmuch as GSA is the client of record for the building.

The walls comprise bays of glass outlined by exposed concrete ledges and limestone-faced fins, and punctuated by exposed concrete columns—the glass bays being interrupted by sections of limestone-veneered masonry.

Masonry walls are insulated to achieve a U value of 0.07. The

double-glazed clear glass occupies roughly only 35 per cent of the wall. To help building energy usage approach GSA's guideline of 55,000 Btu/equivalent gross sq ft/yr, the remaining glass is backed by panels of rigid insulation and gypsum board. On the exposed elevations, Venetian blinds will block unwanted sun. Sliding doors at balconies (identified by the double-thick ledges) are more deeply recessed, and are shielded by the ledges.

The mechanical engineers, using the AXCESS computer energy-use-simulation program, estimate annual energy consumption at about 77,700 Btu/EGSF/yr. If kitchen usage, 13-hr operation, and commercial lighting (the ground floor will be leased to stores) are accounted for, this figure would drop to 57,000.

The architect and consulting engineer Syska & Hennessy collaborated closely in developing systems and architectural elements for delivering air, light and power to the open-plan floor areas as inconspicuously and as efficiently as possible. For example, three air systems are used rather than the usual two; lighting (2¼ W/sq ft) is entirely from the office work stations (task/ambient); and power and communications wiring is run below the 4-in.-high access/raised flooring that is used for all office floors.

Other energy-saving steps by the engineers were: 1) use of vari-

able ventilation and unheated air for garage areas, and 2) supply of only 105 F water to lavatories by a single pipe instead of both cold and 120 F water. Also the building includes a building automation/fire/security system for hvac monitoring and control, fire and sprinkler alarm, automatic smoke purging and security.

The choice site of the building was not easy to acquire. In fact, one of the client's obligations was to rehabilitate the adjacent Winder Building—a five-story cast-iron office structure built in 1842. The plaza between the two buildings has an ice-skating rink that, in summer, is turned into a reflecting pool bridged by duckboards for umbrella-shaded tables. And Sasaki Associates has designed a two-story glass pavilion to abut the Winder building.

Because of two levels of underground parking plus a basement, Lev Zetlin Associates, the structural engineer, had to be careful in their foundation design to minimize underpinning of the Winder Building. Superstructure of the FHLBBB is waffle slab.

FEDERAL HOME LOAN BANK BOARD BUILDING, Washington, D.C. Client: *General Services Administration.* Architects: *Max O. Urbahn Associates, Inc.* Engineers: *Lev Zetlin Associates (structural); Syska & Hennessy, Inc. (mechanical/electrical).* Sitework: *Sasaki Associates.* Interiors: *Max O. Urbahn Associates, Inc.* Contractor: Turner Construction Co.

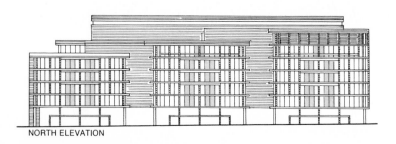

NORTH ELEVATION

Design for modest energy use

Energy consumption is reduced mainly via the wall design, the types of air-handling systems and air distribution systems (right), and task/ambient lighting. Offices in the spine will have full-height partitions (see plan), but with continuous clear-glass transoms to preserve an open feeling. The masonry wall has a 0.07 U value; windows are double-glazed, and those shown with a gray tone in the elevation have insulated panels in back.

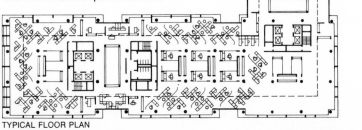

TYPICAL FLOOR PLAN

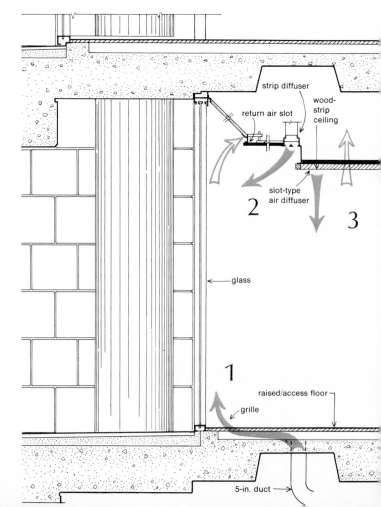

strip diffuser

return air slot

wood-strip ceiling

slot-type air diffuser

2

3

glass

1

raised/access floor

grille

5-in. duct

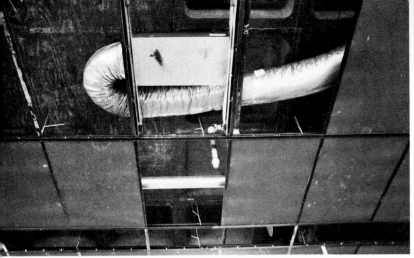

Three air systems for three zones minimize fan energy use

The engineers divided air distribution into three separate systems, each with its own characteristics for doing its job efficiently. The three are illustrated in the section across page, and in the photographs across the top. At the exterior wall, the engineers have used a constant-volume reheat system (reheat coil at air handlers) to produce a curtain of air in front of the glass and walls to neutralize conducted heat or cold, to afford comfort near them. This system need not operate when the outdoor temperature is from 60 to 80 F. Air-return ducting for this system is shown at right. For the second zone, the perimeter (exterior wall to 12-15 ft

inside), the engineers chose a variable-air-volume system that neutralizes the loads of lights, people, and direct solar gain when it occurs. (The linear supply grille for this system is shown across page.) For the third zone, the interior, the engineers decided to use air-powered, above-ceiling induction boxes. When the load drops (fewer lights or people), the system reduces primary air supplied, and thus fan energy. This air is supplied from slot-type diffusers (see above), and is returned through slots between the ceiling wood strips to the plenum. Black-painted acoustical panels over the wood strips close off the plenum.

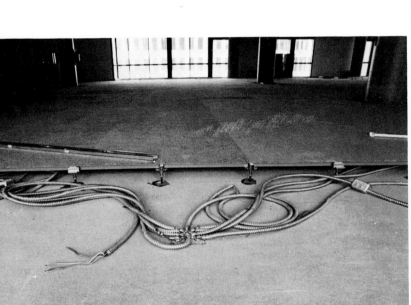

Electrified floor panels allow task lighting to be anywhere

Floors are blanketed with 4-in.-high access flooring that gives complete flexibility for lighting, 120-v power and telephone wires. (Carpet squares were factory applied.) Every 30- by 30-ft area has a junction box buried in the slab. Each of these has the capacity for 10 lengths of flexible metal conduit, though the client is installing only six at this time in most areas. Generally, an electrified access panel will be used every 150 sq ft.

The wood-strip ceiling defines the spine of the building; white acoustical tile, the perimeter. Though the wood

reflectance diminishes potential ambient light, the effect is even, not harsh. The 4-ft "sausage" fixture has two HO-fluorescent lamps for uplight and one HO-lamp for downlight. A refractive-grid, low-brightness lens is on top (the consultant thought occupants should be aware of the source), and a twin-beam lens is on the bottom. The utility module at the corner has deluxe-white-mercury uplight. A flexible-arm incandescent fixture will replace the extendable mock-up fluorescent unit. Lighting by Design Decisions of Syska & Hennessy.

Lighting system in a balconied library

The powerful and sensuous interiors of Toronto's new Metropolitan Library proceed directly from architect Raymond Moriyama's concern that users enjoy freedom of movement and quick comprehension of the library's organization.

Openness extends to the ceiling planes, too, though this is as much functional as it is esthetic.

The ceiling is a series of baffles that not only help shield the lighting fixtures from view but also absorb sound.

But technically, the most significant feature of the ceiling is the variation in luminaire spacing and location to permit variation in lighting levels for different space uses—reading, stacks, offices, conference rooms, etc.

The basic lighting discipline comprimises continuous, single-lamp, fluorescent strip lighting, spaced 7 ft apart; over stacks, however, density is increased to 3½ ft. But a further innovation is the movable, plug-in task luminaire (3-ft, two-lamp), which can be mounted above the ceiling-baffle system where required over a reading desk or work station. Lighting levels will range from about 30 fc, maintained, for the wide spacing to 75 fc for the higher density strips, and spaces under the task luminaires.

Though they were unable to use it in the Scarborough Civic Center, in suburban Toronto, Moriyama and his engineers first considered the ceiling-lighting arrangement for that building, which in its interior spaces has some similarities to the layered-tray interior of the Toronto library. In some ways design goals were similar, too. At Scarborough, Moriyama wanted to encourage contact between residents and borough employees, and to open up the building to the public.

The library, however, is not a borough library, but a Metro library. It is located in a very busy and lively commercial area of Toronto on the edge of downtown that is full of high-rise office buildings, stores, shops, and restaurants. But, more importantly, it is right next to the intersection of the north-south and east-west subway lines.

The architect, wanting to encourage public use and enjoyment of the facility, designed landscaped arcades on two facades of the library, and an interior "street" along which the public may stroll without having to pass through security.

The lighting design, of course, yields energy savings with load averaging out at about 2.2 watts per gross square foot, including a fair amount of incandescent lighting on the main floor, such as the track lighting along the interior "street."

Energy savings also accrue from the well-insulated building shell (polystyrene-insulated wall and roof, U = 0.1, and reflective insulating glass); from the heat recovery system (internal heat extracted by a heat-pump system is sufficient to balance shell losses down to −5 F); from reducing lighting levels and ventilation air at night; and from using one faucet per basin in lavatories with 105 F water.

METROPOLITAN TORONTO LIBRARY. Architect: *Raymond Moriyama, architects and planners.* Engineers: *Robert Halsall & Associates Ltd. (structural); G. Granek & Associates (mechanical); Jack Chisvin & Associates Ltd. (electrical).* Contractors: *The Charles Nolan Company.*

Robert E. Fischer photos

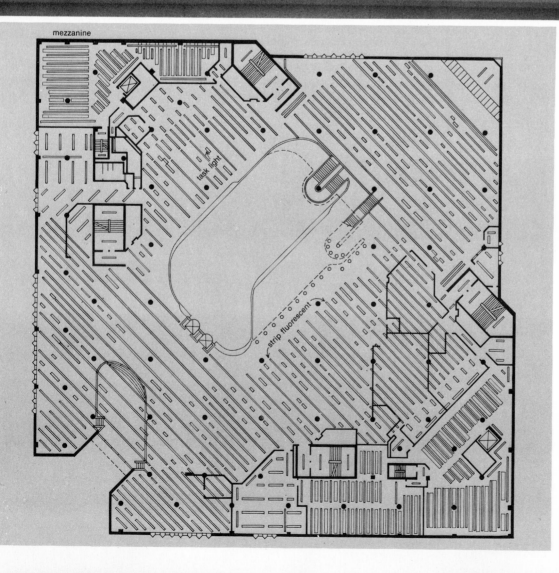

mezzanine

task light

strip fluorescent

The skylight system

A huge, boldly framed skylight is at the top of the atrium, framed with deep, sculptured girders that "play" with the light in its transition to the interior. The girders taper lengthwise, but opposite to one another, and contours change, as well, along their lengths. In addition to reflecting light into the atrium and the outer edges of the balconies, the girders serve esthetically to knit together the two sides of the open well. When supplementary light is desired and at night, mercury downlights in the bottoms of the tapered girders can be turned on.

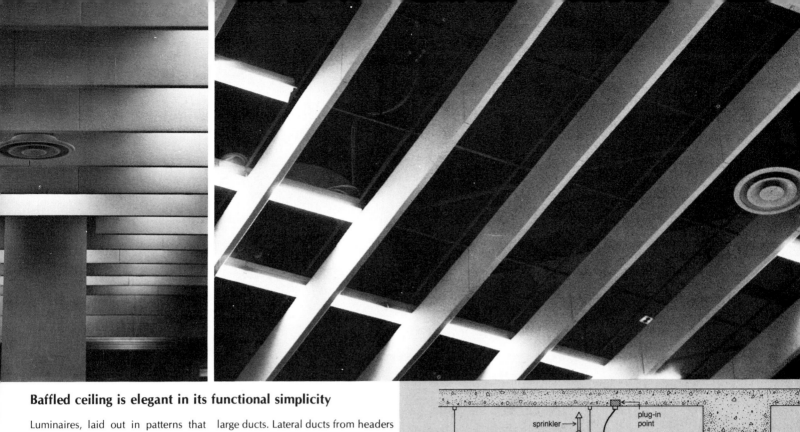

Baffled ceiling is elegant in its functional simplicity

Luminaires, laid out in patterns that suit space-use functions (see plan) produce a subtle play of light and shadow on the ceiling baffles that both shield luminaires from direct view and provide sound absorption. The fabric-covered fiberglass baffles are 18-in. deep, and for reading areas are spaced 36 in. apart. Distance from the floor to the underside of baffles is typically 11 ft 6 in., and distance to the underside of the slab is 19 ft 8 in. This provides ample space for running large ducts. Lateral ducts from headers are connected to circular "elephant trunk" ducts that drop between baffles for air supply. In areas such as office and conference spaces, the ceiling baffles are lowered to 10 ft and spaced closer together at 1½ ft for more sound absorption. The reason for the tall floor-to-floor height was to allow two floors of closed stacks (one a mezzanine) in the corners of floors so there would be sufficient capacity for books the library wishes to control.

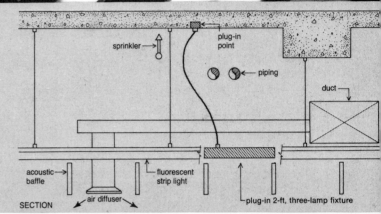

Daylighting
in a low-rise office
building

Robert E. Fischer photos

Architects, it seems, are beginning to rediscover the wonderful qualities of daylight, though with more impressive results, as here, when inventive designers develop new concepts through modern materials and technologies.

John Carl Warneke's office (William Pedersen, designer) created this remarkable daylighted space that has the relaxing ambience of an art museum, yet the greater interest of outdoors because of the modulating effects of its ceiling design elements.

No museum this, however, but rather a 525,000-sq-ft office building (expansible to 1-million) for the Aid Association for Lutherans, the world's largest fraternal insurance organization. The new headquarters sits in the open countryside, amidst farm fields, a few miles from Appleton, Wisconsin, within commuting distance of Oshkosh and Green Bay.

Outwardly, the dramatic ceiling forms on the upper floor of the two-story structure are devices for controlling and suffusing daylight, or supplemental electric light during dark periods from tubular luminaires directly under the reflective-glass skylights.

Because of the balanced brightnesses within the spaces (one never sees the skylights without looking nearly straight up), an observer can hardly believe that lighting levels at midday in

summer are about 200 fc.

The ceiling elements do much more than might be guessed, however. The unique cylindrical shapes in the ceiling, called "socks" by the architect—and now, too, by the building's tour guides—house all the mechanical paraphernalia for delivering and controlling conditioned air, sprinkler piping, roof drainage, and even speakers that create masking sound.

The ceiling also traps sound and absorbs it by means of the "socks" which have fiberglass cores behind the white fabric, stretched in a smooth, taut double-curvature between hoops. The panels are clamped by a special plastic molding.

Because AAL continues to grow, and because support services, such as electronic communications, are changing so rapidly, AAL decided to have a complete access/raised floor for all office areas—nearly 218,000 sq ft.

Mechanical systems and components were selected always with energy savings in mind. Air-handling equipment is in four long, extendable penthouses, running perpendicular to the skylights. Altogether there are 28 air-conditioning systems using packaged air handlers and return-air fans with controllable-pitch blades that are efficient even at low-capacity air delivery.

On the upper level, which has much wider swings in air-conditioning load because of the skylights, three different types of air-handling systems are provided: 1) an interior-zone, variable-air-volume cooling system; 2) a single-zone system for perimeter areas (all glass is double glazed); and 3) a multizone system for areas with large, variable loads.

On the ground floor, perimeter zones are handled similarly to those on the upper floor. Interior zones, however, use air-powered induction boxes, because cooling load stays fairly constant.

Electric heaters are used under the skylights to keep air temperature near them in winter at 72 F to prevent downdrafts. Electric draft-barrier heaters also are provided at all perimeter glass, which turn on at 25 F, and step up to 100 per cent capacity at –20 F.

The all-electric building has three chillers, one of them operat-

ing as an internal-source heat pump. Two 25,000-gal tanks store hot water when it is efficient to do so. The heat pump can be supplemented by two electric boilers.

To implement energy conservation, and also for security and life-safety, the building has a computer-type building automation fire/security system.

The tubular, lensed, twin-lamp luminaires under the skylights can have either half or all lamps switched off as daylight varies. While the control center can provide automatic switching, the owner is doing this manually to learn what is really needed.

AID ASSOCIATION FOR LUTHERANS HEADQUARTERS, Appleton, Wisconsin. Architect: *John Carl Warneke & Associates, Architects and planners.* Engineers: *Paul Weidlinger Associates (structural); Joseph R. Loring & Associates, Inc. (mechanical and electrical).* General contractor: *Oscar J. Boldt Construction Co.*

Two different areas of the AAL building demonstrate the soft, modulated ambience created by the cylindrically-shaped "socks" and deep, sheathed girders that control and disperse daylight from skylights (and supplemental fluorescent fixtures) overhead. The area at top is directly behind a two-story skylit passage on the north side. The area at left is between this passage and an interior garden to the south.

In addition to their light-control function, the fabric-covered ceiling "socks" neatly conceal supply-air ducts and their associated volume-control boxes, and sprinkler piping. Further, they integrate linear diffusers for supply and return air. The "sock" itself serves as the plenum for returning air to fan rooms. A row of fluorescent fixtures used at night or during marginal daylight conditions can be seen above, right. The one luminaire that is lit is on the emergency circuit. The deep rectangular members are sheathed, long-span girders that span as much as 88 ft, and the cross members are roof purlins. While the sky-lighted office space is only on the upper floor, the ground floor, lighted by low-brightness fixtures, shares the two-story-high skylighted passages and interior gardens.

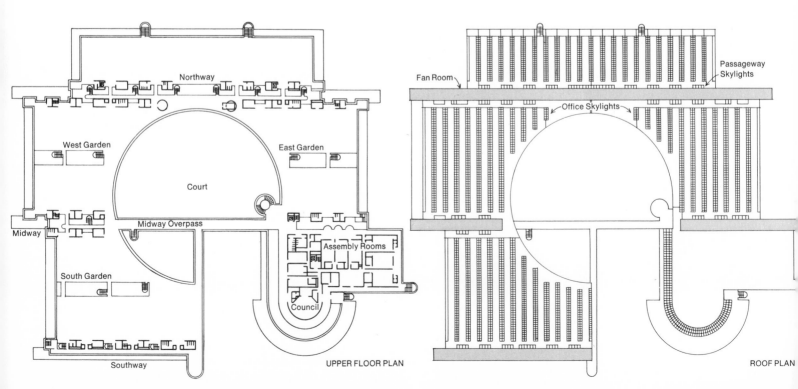

UPPER FLOOR PLAN

ROOF PLAN

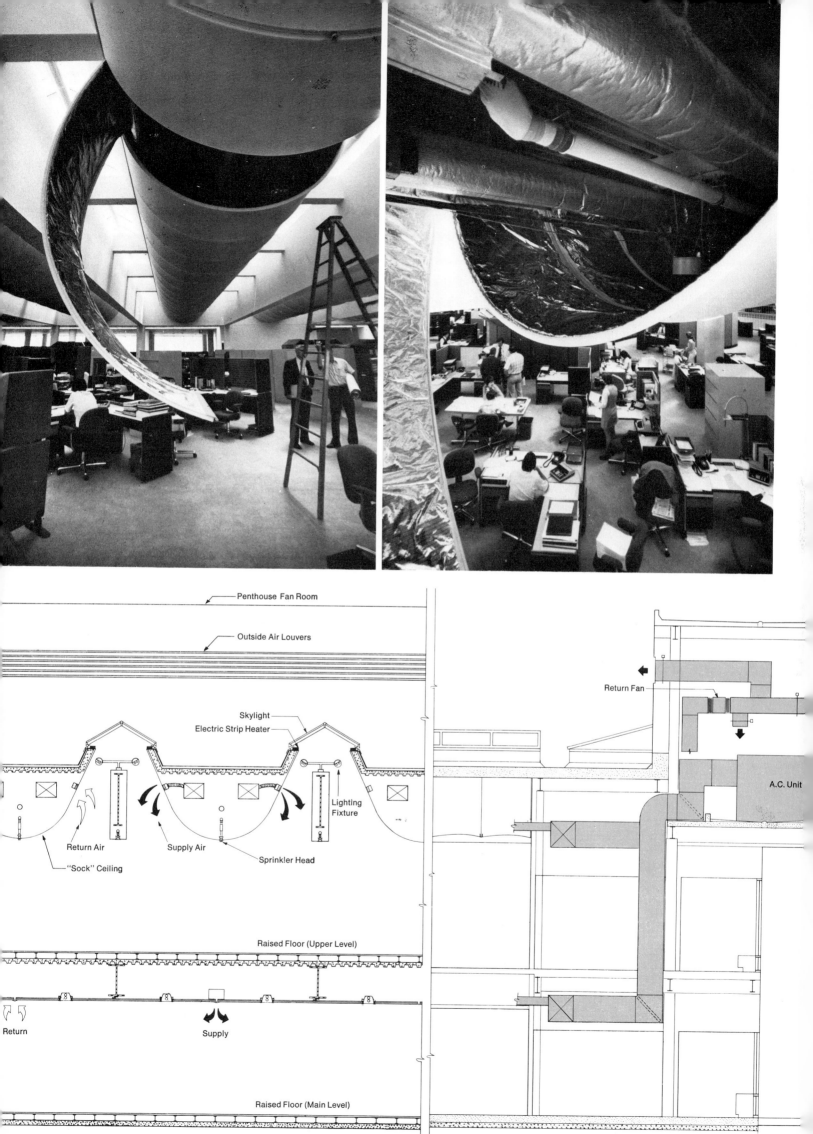

Penthouse Fan Room

Outside Air Louvers

Skylight

Electric Strip Heater

Lighting Fixture

Return Air

Supply Air

Sprinkler Head

"Sock" Ceiling

Raised Floor (Upper Level)

Return

Supply

Raised Floor (Main Level)

Return Fan

A.C. Unit

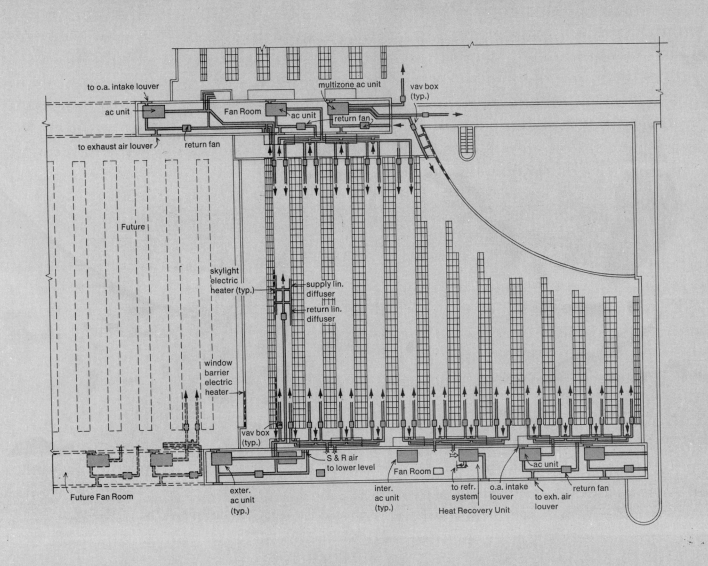

to o.a. intake louver

ac unit

Fan Room

ac unit

multizone ac unit

vav box (typ.)

to exhaust air louver

return fan

return fan

Future

skylight electric heater (typ.)

supply lin. diffuser

return lin. diffuser

window barrier electric heater

vav box (typ.)

S & R air to lower level

Fan Room

Future Fan Room

exter. ac unit (typ.)

inter. ac unit (typ.)

Heat Recovery Unit

to refr. system

o.a. intake louver

to exh. air louver

ac unit

return fan

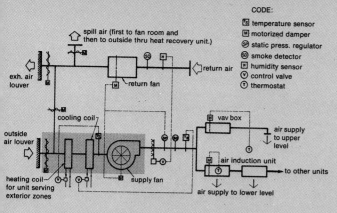

CODE:
- **TS** temperature sensor
- **M** motorized damper
- **SP** static press. regulator
- **SD** smoke detector
- **H** humidity sensor
- **V** control valve
- **T** thermostat

spill air (first to fan room and then to outside thru heat recovery unit.)

return air

exh. air louver

return fan

cooling coil

outside air louver

heating coil for unit serving exterior zones

supply fan

vav box
air supply to upper level

air induction unit
to other units

air supply to lower level

CONTROL DIAGRAM—SINGLE ZONE AC SYSTEM, INTERIOR AND EXTERIOR

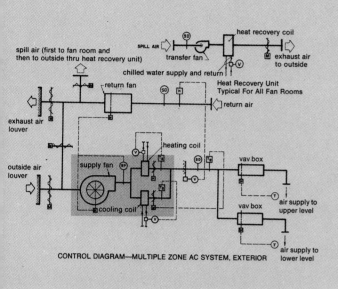

spill air (first to fan room and then to outside thru heat recovery unit)

SPILL AIR

transfer fan

heat recovery coil

exhaust air to outside

chilled water supply and return

Heat Recovery Unit Typical For All Fan Rooms

return fan

return air

exhaust air louver

heating coil

outside air louver

supply fan

cooling coil

vav box
air supply to upper level

vav box
air supply to lower level

CONTROL DIAGRAM—MULTIPLE ZONE AC SYSTEM, EXTERIOR

All air handlers are in linear penthouses that are a design feature of the AAL building. Basically two different types of air handlers (see diagrams) were used to serve three types of zones: 1) a single-zone air conditioner for interior zones, which, with a heating coil added to it, also is used for perimeter zones at the outside faces of the building; 2) a multizone air conditioner with hot and cold decks and damper-mixed air to handle large and varying heating and cooling loads for zones bordering the glass of the circular court, and for the glazed overpass. The chiller with a double-bundle condenser that acts as an internal source heat pump is at top. One of the packaged air handlers is above, right, and one of the variable-pitch-blade return-air fans is directly above.

Electrified panels of the access floor system bring power, telephone and communications to work stations. Now the building has 1200 such panels. A junction box occurs every 400 sq ft, with 6 to 8 flexible metal cables. The panels are fastened to the post supports, and also to each other for rigidity. They are covered with carpet squares kept in place by a magnetic coating on the backs.

Ceiling coffers support luminaires and air diffusers

For the corporate headquarters of Air Products & Chemicals, Inc., near Allentown, Pennsylvania, the design architect, Bernward Kurtz of The Eggers Group, wanted a sculptural feeling in the ceiling design without calling attention to the luminaires.

The solution, developed by Kurtz with his lighting consultant, James D. Kaloudis of Meyer, Strong & Jones, consulting engineers, is special pyramidal coffers supporting low-brightness, parabolic-type, twin-beam luminaires. And because partitions for some parts of the building could be as thick as 6 in. for fire protection purposes, Kurtz wanted ceiling runners 6-in. wide also, so that partitions would not interfere with the slope of the coffers. Since runners 6-in. wide were not available as stock items in pressed-steel sections, Kurtz found a fabricator to manufacture them, and he had the exposed 6-in.-wide strips finished in different colors to designate different floors. He also had the fabricator produce the metal framing and end panels for the coffer he designed—striking for its simplicity, and the ingenuity dis-

Don Fraser, Academy Studios Ltd.

played—to support the louvered parabolic luminaire and air diffuser saddle on top of the luminaire—scarcely a lightweight item.

The owner, represented by engineer Philip F. Newman, now manager of general services for Air Products & Chemicals, made clear that the building should be not only flexible and efficient in the management of space, but also efficient in its use of energy. Flexibility for receptacle and communication wiring is provided by cellular flooring with in-floor fixtures, and H-cuts in the carpet to conceal the fixtures while allowing wiring egress.

The building consumes significantly less energy than similar existing buildings, according to John E. Plantinga, partner-in-charge for Meyer, Strong & Jones, even though it has some large areas with high heat loads and high fresh air requirements, such as the computer area, kitchen complex, cafeteria and print shop. Energy is saved by the lighting system (only 1.8 watts per sq ft), by the building envelope (precast panels are insulated, windows are double-glazed and take up only 25 per cent of the exterior), and by the hvac system (a variable-primary-air type that utilizes enthalpy control for optimizing free cooling via outdoor air). Electrical demand control also is provided as part of the building automation system.

Don Fraser, Academy Studios Ltd.

A 380,000-sq-ft corporate headquarters building, added to an existing campus of office, research and industrial buildings, for Air Products & Chemicals, Inc., in Trexlertown, Pennsylvania, has garnered several awards for its energy-conserving features. The headquarters is a series of four building units in a zigzag configuration. Exterior is insulated precast concrete panels and bronze-tinted, double-glazed windows. For most of the building, single-lamp luminaires were employed, using only 1.8 watts per sq ft, and providing about 75-80 footcandles in large open spaces. In several small partitioned interior spaces used by executives' secretaries, a single two-lamp luminaire was installed above the desks. In the drafting room, two-lamp troffers (one lamp above the other) alternate with single-lamp troffers to give more light. Ceiling runners suspend smoke detectors in the computer area.

Lawrence S. Williams

The coffer frame shown at the bottom of the page supports the luminaire and its associated air supply/return saddle. The frame for the luminaire is supported by the metal end panels together with the two stiffener pieces and angles shown. The metal panel is bent for rigidity and to provide a right angle section at the bottom to be received by the main runner, and also to provide a flange at the top to support lay-in acoustical panels. The 55-in. cross runners fit between the 10-ft main runners.

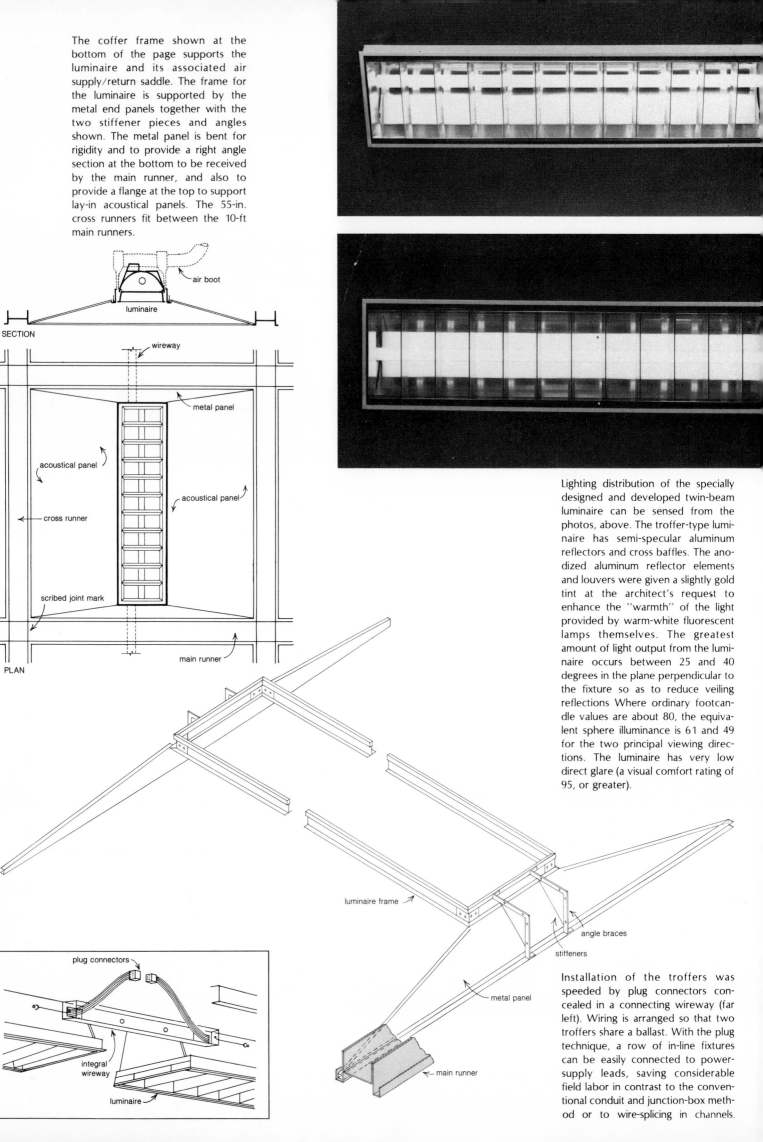

SECTION

air boot

luminaire

wireway

metal panel

acoustical panel

acoustical panel

cross runner

scribed joint mark

PLAN

main runner

Lighting distribution of the specially designed and developed twin-beam luminaire can be sensed from the photos, above. The troffer-type luminaire has semi-specular aluminum reflectors and cross baffles. The anodized aluminum reflector elements and louvers were given a slightly gold tint at the architect's request to enhance the ''warmth'' of the light provided by warm-white fluorescent lamps themselves. The greatest amount of light output from the luminaire occurs between 25 and 40 degrees in the plane perpendicular to the fixture so as to reduce veiling reflections Where ordinary footcandle values are about 80, the equivalent sphere illuminance is 61 and 49 for the two principal viewing directions. The luminaire has very low direct glare (a visual comfort rating of 95, or greater).

luminaire frame

angle braces

stiffeners

metal panel

plug connectors

integral wireway

luminaire

main runner

Installation of the troffers was speeded by plug connectors concealed in a connecting wireway (far left). Wiring is arranged so that two troffers share a ballast. With the plug technique, a row of in-line fixtures can be easily connected to power-supply leads, saving considerable field labor in contrast to the conventional conduit and junction-box method or to wire-splicing in channels.

AIR-CONDITIONING SYSTEM SCHEMATIC

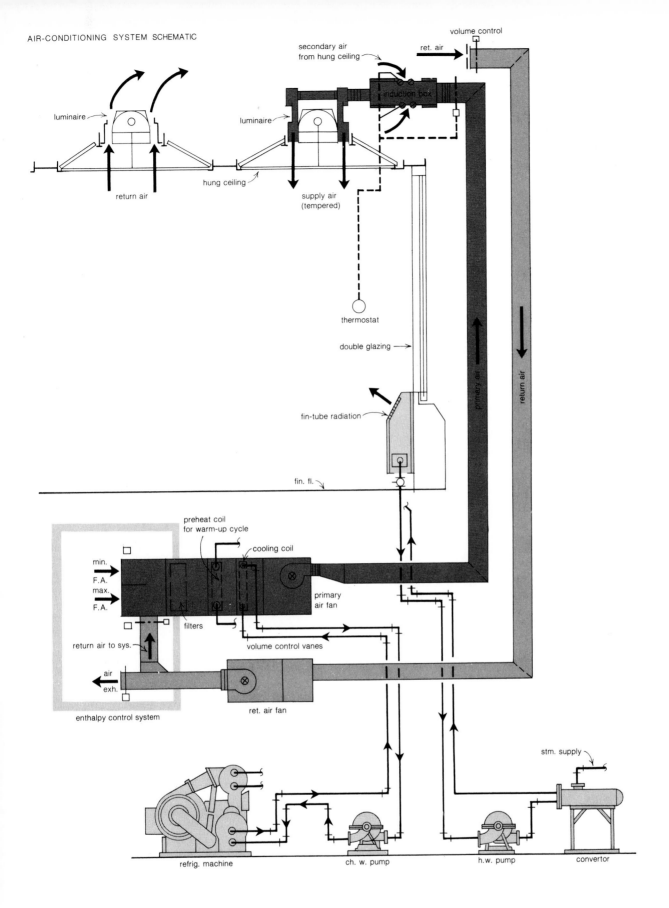

Spaces are conditioned by means of a variable-volume system. Supply air to spaces is constant, while the ratio of primary (cooled) air to return air is varied by the above-ceiling induction boxes in response to the room thermostats. The only source for tempering is the return air from the ceiling plenum. Primary air can be throttled to 40 per cent of maximum. Perimeter offices are heated by fin-tube radiation. Air is supplied to rooms and returned to the ceiling plenum through slots at the sides of the troffers. An enthalpy control system senses the temperature and humidity of outdoor air and "decides" when it can be used for "free" cooling. Because of the low energy taken by the lighting, the engineers were able to reduce the refrigeration capacity, air distribution and electrical distribution. A study on the cost implications of design features and on energy-system strategies, done for the client by consultant L. G. Spielvogel, concluded that the building generally meets the energy budget guidelines suggested by Federal agencies.

Wall washers illuminate library work space

In their project for unpartitioned library space at New York State University's Agricultural & Technical College in Morrisville, architects Morris Ketchum, Jr. & Associates resisted the use of a flat ceiling with flush lighting fixtures in order to avoid the nervous pattern this arrangement so often produces. They provided instead a stronger rhythm established by large columns supporting deep oblong coffers. They then asked electrical engineer Henry Wald and his lighting associate James Kaloudis to devise a structurally integrated system that would provide soft but adequate general illumination and that would also define architectural spaces.

About the only place to position lighting fixtures without mutilating the ceiling's appearance was around the upper edges of the coffers. Conventional strip lighting, however, would not only waste a good deal of the available illumination against the sides of the coffers, which would obstruct its distribution into adjacent areas; it would fail the essential task of providing sufficient light for readers and librarians working in the center of the areas beneath each coffer. Paradoxically, what was needed to obtain evenly spread over-all light was a strongly directional beam to bring light to the center from all four sides of the coffer.

The strip fixtures that provide this beam are standard wall-washer lenses ingeniously tilted 20 deg. Rather than directing the beam away from the wall at about 15 deg, as it does when normally installed, the angled lens projects its main beam at about 35 deg, sufficient to place light where it is needed. As a lighting bonus, 35 deg is also a favorable angle with respect to reflected glare.

Despite the len's relatively restricted area of distribution—its spread is only 20-25 deg—a small amount of backlight remains to wash the sides of the coffer. Besides emphasizing the ceiling form, this wash also contributes importantly to visual comfort in the space. One of the designers' early concerns was that the lens, which was obviously never intended to be looked at directly, would be unacceptably brilliant when tilted and highly visible. The engineers ran subjective tests, viewing the angled lens alone and beside a light-colored vertical surface. The visual effect of this surface—in place, the lighted sides of the coffers—is to offer the eye an expanse of brightness that diminishes awareness of the still brighter strip of the lens.

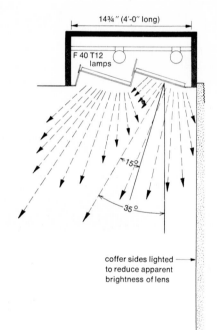

14¾" (4'-0" long)

F 40 T12 lamps

15°

35°

coffer sides lighted to reduce apparent brightness of lens

Architects: Morris Ketchum, Jr. & Associates. Engineers: Wald & Zigas (mechanical/electrical); Ames & Selnick (structural).

carrels and stacks

carrels and stacks   lobby

reading and stacks

Flexible wiring systems:
a catalog of current technology

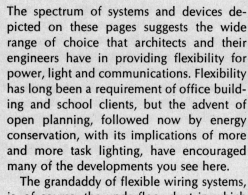

The spectrum of systems and devices depicted on these pages suggests the wide range of choice that architects and their engineers have in providing flexibility for power, light and communications. Flexibility has long been a requirement of office building and school clients, but the advent of open planning, followed now by energy conservation, with its implications of more and more task lighting, have encouraged many of the developments you see here.

The grandaddy of flexible wiring systems, is, of course, the underfloor duct in which raceways are embedded in concrete fill. This was followed by cellular floor systems in which the cells are integral with the structural metal deck. Cellular floors have gone through a number of permutations, including the use of large-type cells for air distribution. More recently, attention has been paid to cosmetics of the outlets, some of which can be fully concealed by carpet.

When the floor construction is concrete instead of steel decking, then the alternative system to poke-through (which requires coring slabs and using rated fittings) is an overhead raceway with associated power/telephone poles. Code allows these raceways in air-handling ceilings so long as the ceiling is accessible. Originally, power poles were hard-wired to raceways, but then power plugs and raceway receptacles were developed that made the power poles much more flexible.

If lighting flexibility alone is sought, then cable sets either with or without overhead raceways may be used.

Youngest of all in the family of flexible wiring systems is the use of access flooring for power and communications distribution. From the aspect of flexibility, the advantages are obvious. Insurance companies and banks have been early users, partly because of their growing dependence upon computer terminals, video displays, and other electronic and communication equipment.

Light track is an exposed raceway that architects have grown fond of because flexibility is limited only by track locations.

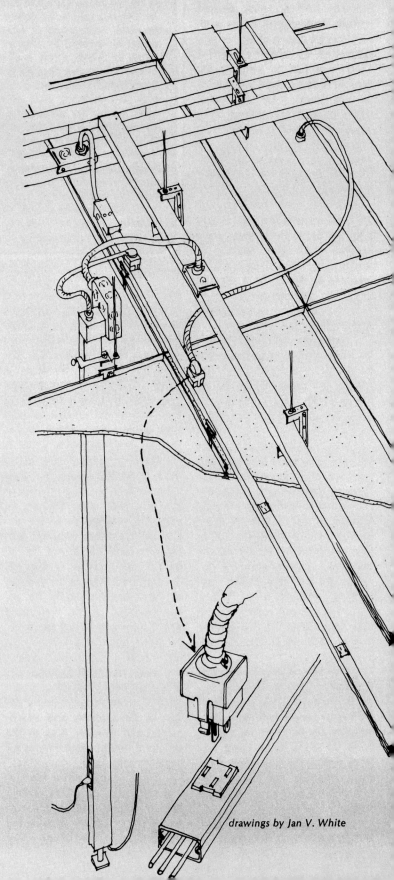

drawings by Jan V. White

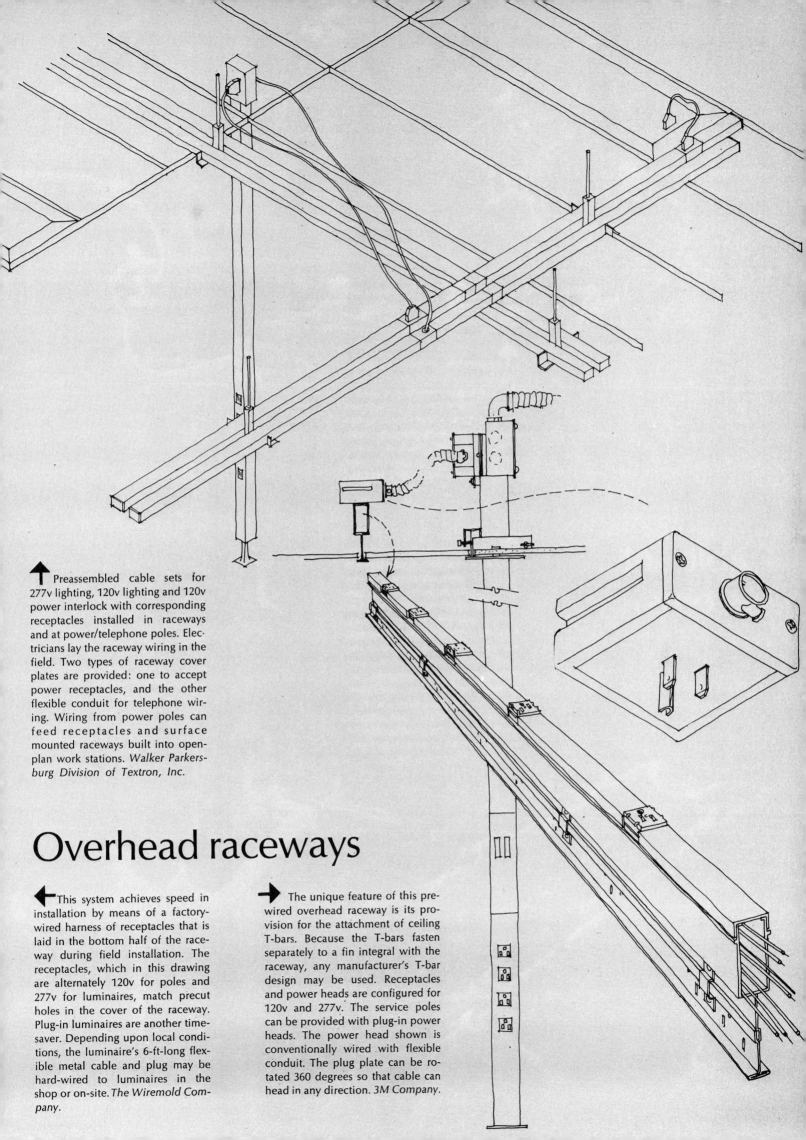

▲ Preassembled cable sets for 277v lighting, 120v lighting and 120v power interlock with corresponding receptacles installed in raceways and at power/telephone poles. Electricians lay the raceway wiring in the field. Two types of raceway cover plates are provided: one to accept power receptacles, and the other flexible conduit for telephone wiring. Wiring from power poles can feed receptacles and surface mounted raceways built into open-plan work stations. *Walker Parkersburg Division of Textron, Inc.*

Overhead raceways

◄ This system achieves speed in installation by means of a factory-wired harness of receptacles that is laid in the bottom half of the raceway during field installation. The receptacles, which in this drawing are alternately 120v for poles and 277v for luminaires, match precut holes in the cover of the raceway. Plug-in luminaires are another time-saver. Depending upon local conditions, the luminaire's 6-ft-long flexible metal cable and plug may be hard-wired to luminaires in the shop or on-site. *The Wiremold Company.*

▶ The unique feature of this pre-wired overhead raceway is its provision for the attachment of ceiling T-bars. Because the T-bars fasten separately to a fin integral with the raceway, any manufacturer's T-bar design may be used. Receptacles and power heads are configured for 120v and 277v. The service poles can be provided with plug-in power heads. The power head shown is conventionally wired with flexible conduit. The plug plate can be rotated 360 degrees so that cable can head in any direction. *3M Company.*

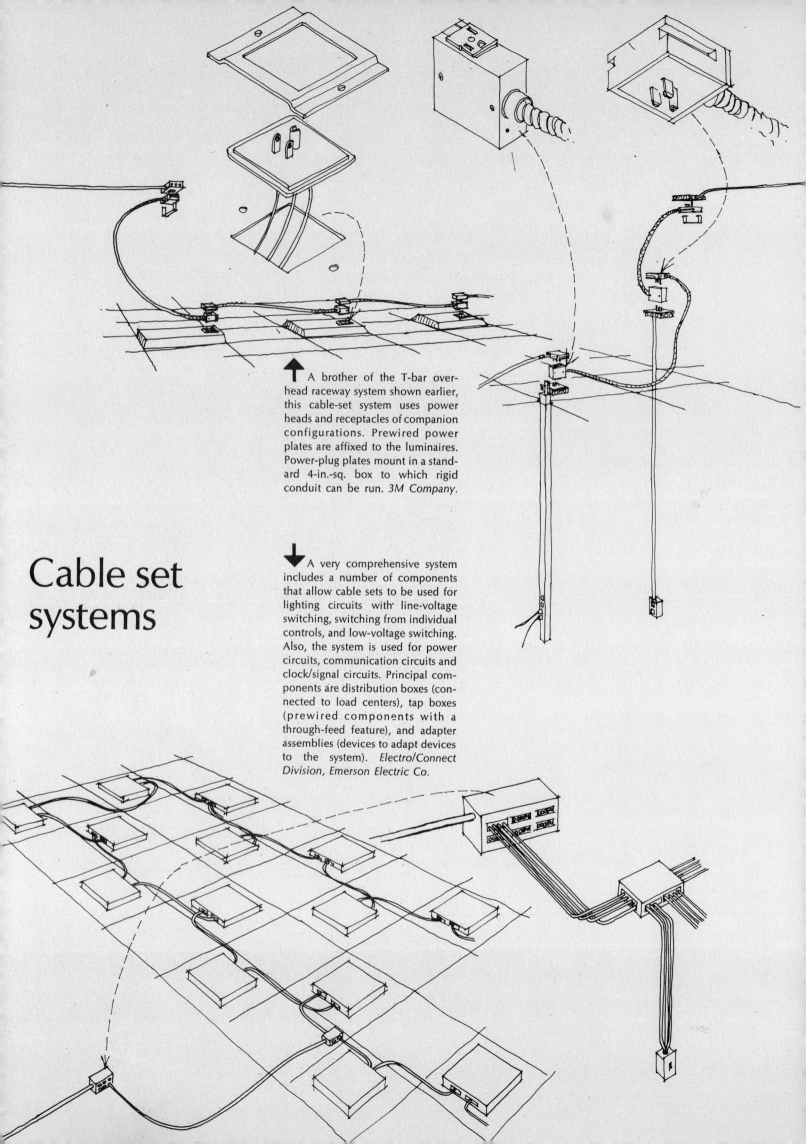

Cable set systems

↑ A brother of the T-bar overhead raceway system shown earlier, this cable-set system uses power heads and receptacles of companion configurations. Prewired power plates are affixed to the luminaires. Power-plug plates mount in a standard 4-in.-sq. box to which rigid conduit can be run. *3M Company*.

↓ A very comprehensive system includes a number of components that allow cable sets to be used for lighting circuits with line-voltage switching, switching from individual controls, and low-voltage switching. Also, the system is used for power circuits, communication circuits and clock/signal circuits. Principal components are distribution boxes (connected to load centers), tap boxes (prewired components with a through-feed feature), and adapter assemblies (devices to adapt devices to the system). *Electro/Connect Division, Emerson Electric Co.*

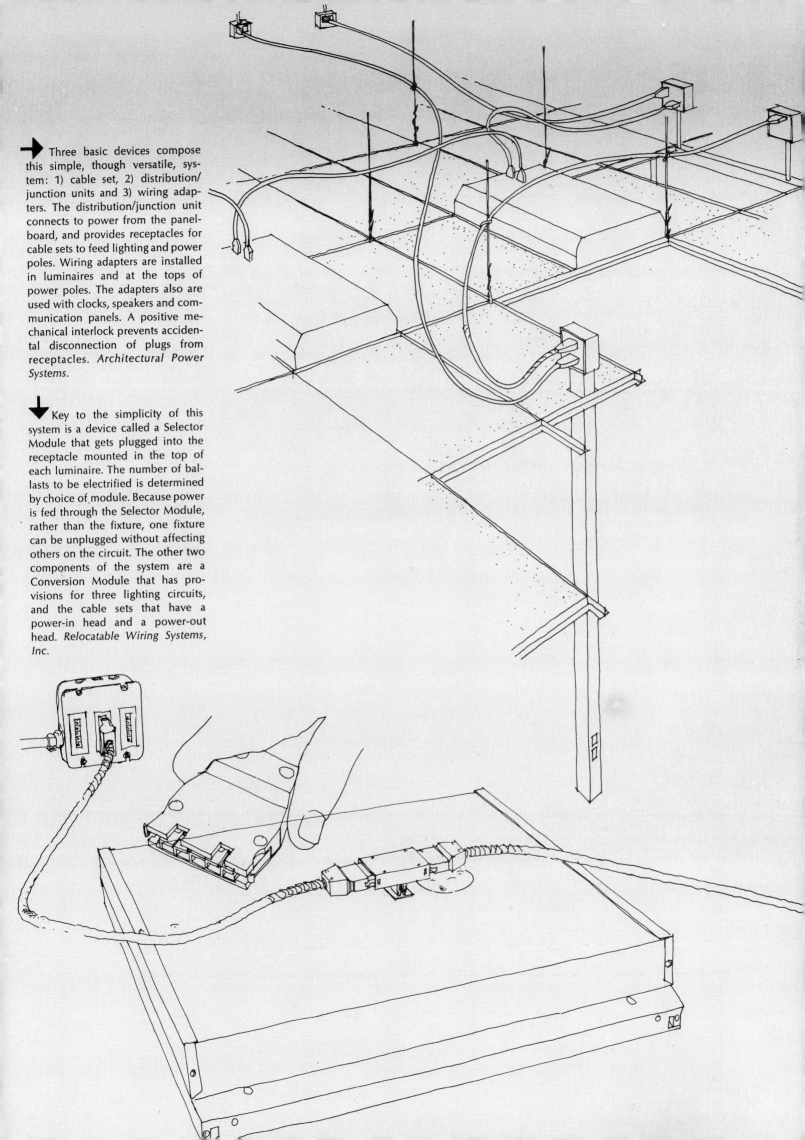

▶ Three basic devices compose this simple, though versatile, system: 1) cable set, 2) distribution/junction units and 3) wiring adapters. The distribution/junction unit connects to power from the panelboard, and provides receptacles for cable sets to feed lighting and power poles. Wiring adapters are installed in luminaires and at the tops of power poles. The adapters also are used with clocks, speakers and communication panels. A positive mechanical interlock prevents accidental disconnection of plugs from receptacles. *Architectural Power Systems.*

▼ Key to the simplicity of this system is a device called a Selector Module that gets plugged into the receptacle mounted in the top of each luminaire. The number of ballasts to be electrified is determined by choice of module. Because power is fed through the Selector Module, rather than the fixture, one fixture can be unplugged without affecting others on the circuit. The other two components of the system are a Conversion Module that has provisions for three lighting circuits, and the cable sets that have a power-in head and a power-out head. *Relocatable Wiring Systems, Inc.*

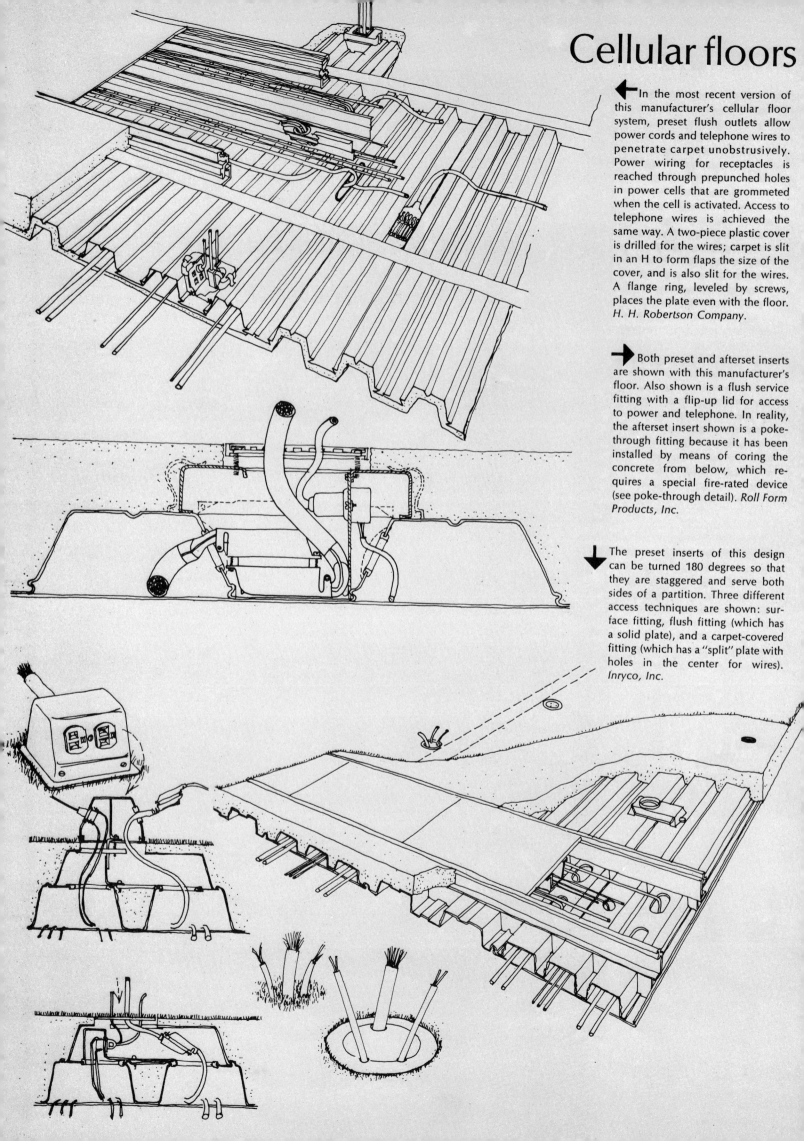

Cellular floors

◀ In the most recent version of this manufacturer's cellular floor system, preset flush outlets allow power cords and telephone wires to penetrate carpet unobstrusively. Power wiring for receptacles is reached through prepunched holes in power cells that are grommeted when the cell is activated. Access to telephone wires is achieved the same way. A two-piece plastic cover is drilled for the wires; carpet is slit in an H to form flaps the size of the cover, and is also slit for the wires. A flange ring, leveled by screws, places the plate even with the floor. *H. H. Robertson Company.*

▶ Both preset and afterset inserts are shown with this manufacturer's floor. Also shown is a flush service fitting with a flip-up lid for access to power and telephone. In reality, the afterset insert shown is a poke-through fitting because it has been installed by means of coring the concrete from below, which requires a special fire-rated device (see poke-through detail). *Roll Form Products, Inc.*

▼ The preset inserts of this design can be turned 180 degrees so that they are staggered and serve both sides of a partition. Three different access techniques are shown: surface fitting, flush fitting (which has a solid plate), and a carpet-covered fitting (which has a "split" plate with holes in the center for wires). *Inryco, Inc.*

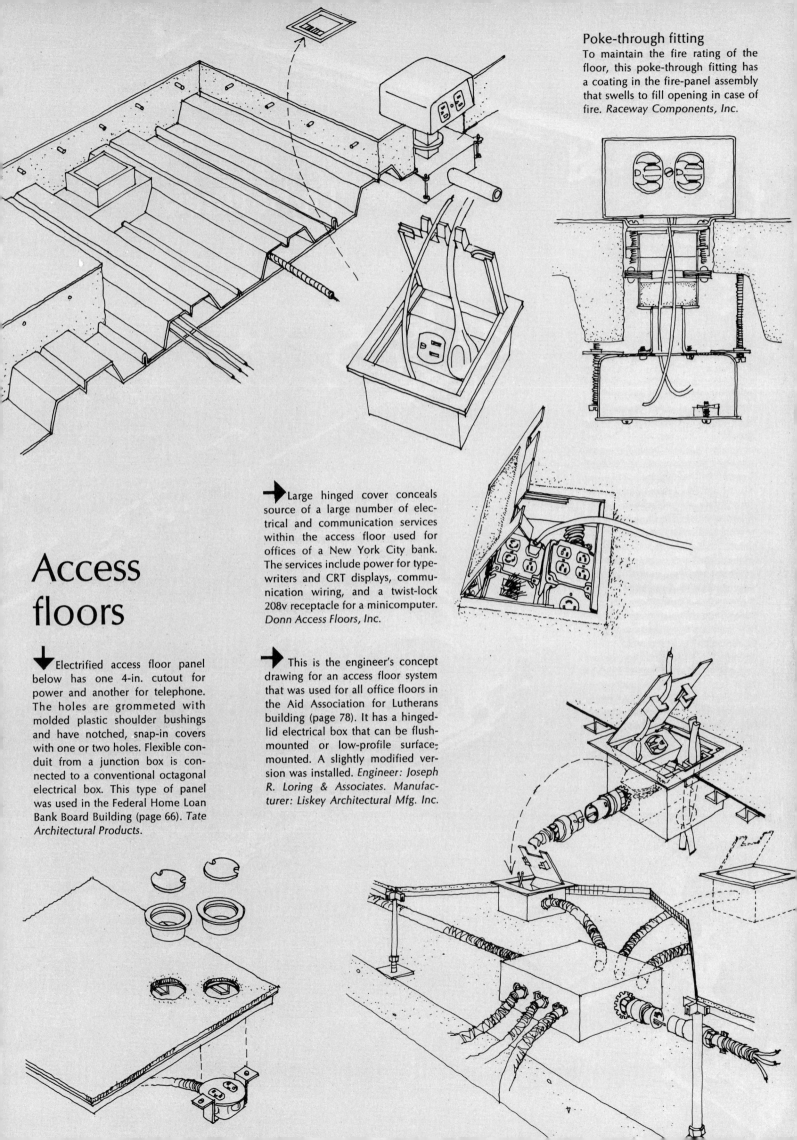

Access floors

▼Electrified access floor panel below has one 4-in. cutout for power and another for telephone. The holes are grommeted with molded plastic shoulder bushings and have notched, snap-in covers with one or two holes. Flexible conduit from a junction box is connected to a conventional octagonal electrical box. This type of panel was used in the Federal Home Loan Bank Board Building (page 66). *Tate Architectural Products.*

▶Large hinged cover conceals source of a large number of electrical and communication services within the access floor used for offices of a New York City bank. The services include power for typewriters and CRT displays, communication wiring, and a twist-lock 208v receptacle for a minicomputer. *Donn Access Floors, Inc.*

▶This is the engineer's concept drawing for an access floor system that was used for all office floors in the Aid Association for Lutherans building (page 78). It has a hinged-lid electrical box that can be flush-mounted or low-profile surface-mounted. A slightly modified version was installed. *Engineer: Joseph R. Loring & Associates. Manufacturer: Liskey Architectural Mfg. Inc.*

Poke-through fitting

To maintain the fire rating of the floor, this poke-through fitting has a coating in the fire-panel assembly that swells to fill opening in case of fire. *Raceway Components, Inc.*

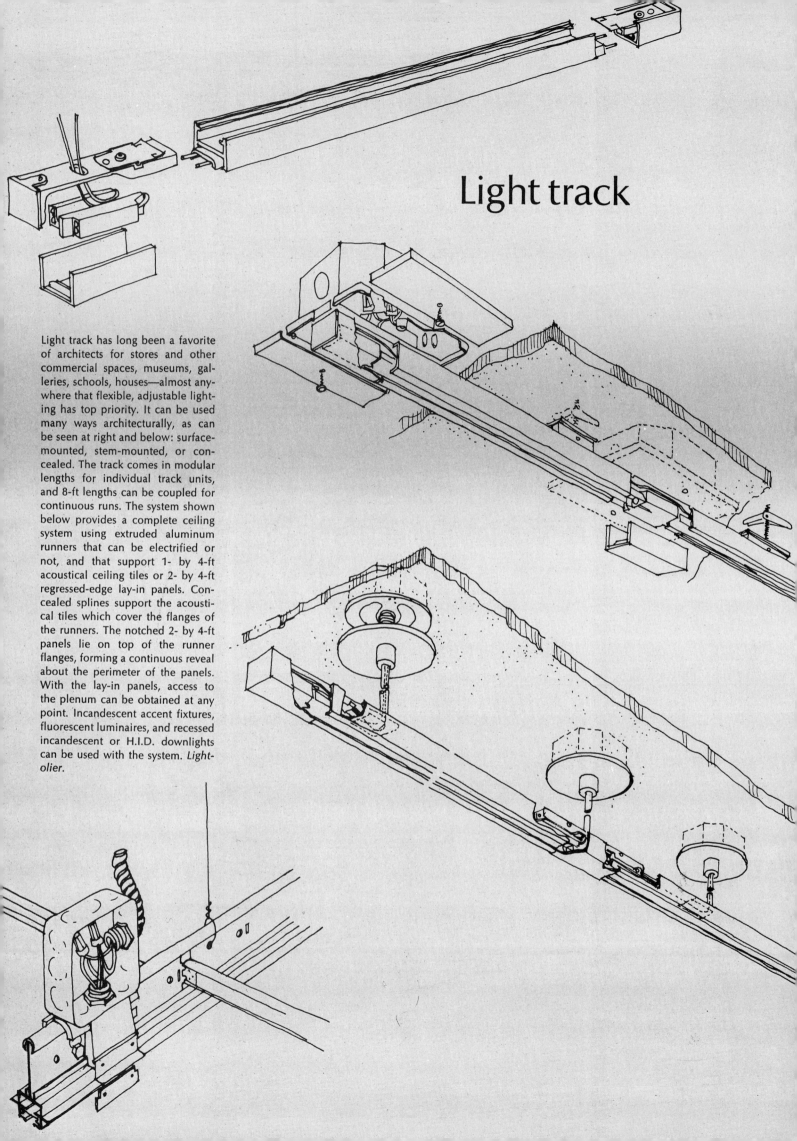

Light track

Light track has long been a favorite of architects for stores and other commercial spaces, museums, galleries, schools, houses—almost anywhere that flexible, adjustable lighting has top priority. It can be used many ways architecturally, as can be seen at right and below: surface-mounted, stem-mounted, or concealed. The track comes in modular lengths for individual track units, and 8-ft lengths can be coupled for continuous runs. The system shown below provides a complete ceiling system using extruded aluminum runners that can be electrified or not, and that support 1- by 4-ft acoustical ceiling tiles or 2- by 4-ft regressed-edge lay-in panels. Concealed splines support the acoustical tiles which cover the flanges of the runners. The notched 2- by 4-ft panels lie on top of the runner flanges, forming a continuous reveal about the perimeter of the panels. With the lay-in panels, access to the plenum can be obtained at any point. Incandescent accent fixtures, fluorescent luminaires, and recessed incandescent or H.I.D. downlights can be used with the system. *Lightolier.*